ENVIRONMENTAL REMEDIATION TECHNOLOGIES, REGULATIONS AND SAFETY

COAL FLY ASH

PROPERTIES, APPLICATIONS AND PERFORMANCE

ENVIRONMENTAL REMEDIATION TECHNOLOGIES, REGULATIONS AND SAFETY

Additional books and e-books in this series can be found on Nova's website under the Series tab.

Environmental Remediation Technologies, Regulations and Safety

Coal Fly Ash

Properties, Applications and Performance

Miriam Gray
Editor

nova
science publishers
New York

Library of Congress Cataloging-in-Publication Data

Names: Gray, Miriam (Materials researcher), editor.
Title: Coal fly ash : properties, applications and performance / editor, Miriam Gray.
Other titles: Coal fly ash (Nova Science Publishers)
Description: Hauppauge, New York : Nova Science Publishers, Inc., [2018] | Series: Environmental remediation technologies, regulations and safety | Includes index.
Identifiers: LCCN 2018044355 (print) | LCCN 2018046377 (ebook) | ISBN 9781536145120 (ebook) | ISBN 9781536145113 (softcover) | ISBN 9781536145120 (ebook)
Subjects: LCSH: Fly ash--Recycling. | Portland cement--Additives. | Asphalt emulsion mixtures. | Chemical tests and reagents.
Classification: LCC TA455.F55 (ebook) | LCC TA455.F55 C63 2018 (print) | DDC 620.1/98--dc23
LC record available at https://lccn.loc.gov/2018044355

Published by Nova Science Publishers, Inc. † New York

CONTENTS

PREFACE

Due to rising energy demand, the amount of coal utilization is escalating at a fast rate, leading to a rise in ash generation. Coal Fly Ash: Properties, Applications and Performance begins by presenting a study wherein the physiochemical properties of coal ash were analysed and based on that, their potential for reuse was identified.

Next, the synergy between Portland cement and coal fly ash was investigated. Raw materials were characterized in terms of chemical and mineralogical composition and physical properties. Mortars were manufactured and the effect of the addition of different ratios of coal fly ash as a partial substitute of Portland cement were studied.

Following this, an investigation was carried out to explore the behaviour of coal ash with hydrated lime as a mineral filler in hot mix asphalt, in an attempt to determine a sustainable solution for the management of coal ash in Mauritius.

Finally, coal fly ash is a low cost source of both silica and alumina. The closing chapter describes the synthesis process used to obtain zeolite Na-A from coal fly ash via fusion with NaOH followed by hydrothermal reaction.

Chapter 1 - Owing to the rise in energy demand, the amount of coal utilized is escalating at a fast rate causing an increase in the amount of ash being generated. These ash are mainly being disposed of in ash ponds/pits or at the landfill posing a threat to human health and the environment. In this study, the physico-chemical properties of coal ash were analysed and based on that, their potential reuse were identified. The two methods implemented for this study were mainly the leaching method to reduce the level of heavy metals to the permissible limit and thermal treatment to reduce the amount of carbon to the required level. The basic laboratory test performed were moisture content, pH and loss on ignition. The leaching behaviour of phosphate was also analysed through variation of acid concentration and time. Also, a set-up of an ash filter bed on a lab-scale basis was modelled for treatment of dye wastewater where parameters such as COD and colour of the untreated dye wastewater and treated dye wastewater were tested and further compared. From the different interpretations obtained, the potential reuse of ash was identified.

Chapter 2 - Recycling is a strategy to consume the waste of several industries as raw material. Coal-fired power plants produce millions of tons of fly ash that cause environmental problems. In addition, these waste materials need large deposit areas with high cost. In this study, the synergy between Portland cement and coal fly ash was investigated. Raw materials were characterized in terms of chemical and mineralogical composition and physical properties. Mortars were manufactured and the effect of the addition of different ratios of coal fly ash (0-80 wt. %) as a partial substitute of Portland cement was studied. The mineralogical composition of the sustainable mortars has been analyzed by XRD and FT-IR. Physical and mechanical properties as bulk density, water absorption by capillarity, flexural and compressive strength of hardener mortars have been determined according to EN standards. The results show that the bulk density of the cement mortars is slightly affected by the coal fly ash addition but leads to an increase of the water absorption by capillarity after of 7 and 28 of curing time. The densification of fly coal ash mortar during the curing age contributed to a significant reduction of the porosity and the water absorption by capillarity. The flexural and compressive strength decrease as the ratio of coal fly ash increases with respect to the control mortar, except for the mortar that incorporates 20 wt % of coal fly ash. The mortars containing up to 20 wt % of coal fly ash offer an opportunity to recycle wastes and present physical and mechanical properties similar to control mortars, obtaining economical and environmental benefits.

Chapter 3 - An investigation has been carried out in this study to explore the behaviour of using coal ash with hydrated lime as mineral filler in hot mix asphalt, in an attempt to a sustainable solution for the management of coal ash in Mauritius. In this study, the fly ash and bottom ash were incorporated in various percentages (0%, 1%, 2%, 3% and 4%) each with a fix percentage (1%) of hydrated lime as mineral fillers and as partial replacement of fine aggregate in 40/50 grade bitumen. The asphalt concrete mixes were prepared for a fix asphalt content (5.7%) hence were tested to evaluate their engineering properties which encompass moisture susceptibility, rutting and temperature susceptibility. Even environmental impacts while being in asphalt was evaluated. Mixes modified with the fillers were found to have improved temperature susceptibility and rutting, also showed lower moisture susceptibility and the leachates detected were within acceptable range. It is therefore concluded that both fly ash and bottom ash can be added successfully in asphalt up to 4% each with 1% hydrated lime to produce an asphalt concrete with better serviceability while being safe to the environment.

Chapter 4 - This chapter describes the synthesis process to obtain zeolite Na-A from coal fly ash via fusion with NaOH followed by hydrothermal reaction. Coal fly ash is a low cost source of both silica and alumina. The molar ratio of SiO_2/Al_2O_3 for zeolite 4A formation was achieved by the addition of sodium aluminate solution or aluminum waste from tertiary industry as aluminum sources. The effects of temperature and time at the fusion step and the hydrothermal reaction time on crystalline products were investigated. The synthesized products were studied for their purity and yield as well as were characterized by different techniques such as X-ray powder diffraction (XRD), scanning electron microscopy (SEM) and cation exchange capacity (CEC), among others. Analysis of the properties of the raw materials and the products demonstrates that the by-products can be used to produce a high purity zeolite A. The optimum conditions related to the synthesis was set and the samples were pure, highly crystalline and white. The synthesized zeolites were used for both the removal of Cd (II) and Zn (II) from aqueous solution. The results showed that coal fly ash and aluminum source can be transformed into zeolite A, which is considered as value-added material with adsorption properties.

In: Coal Fly Ash
Editor: Miriam Gray
ISBN: 978-1-53614-511-3

Chapter 1

A GENERAL OVERVIEW OF COAL FLY ASH

Shoobhashini Keesoondoyal
and Sindra L. Summoogum-Utchanah*
Department of Chemical and Environmental Engineering,
Faculty of Engineering, University of Mauritius,
Reduit, Mauritius

ABSTRACT

Owing to the rise in energy demand, the amount of coal utilized is escalating at a fast rate causing an increase in the amount of ash being generated. These ash are mainly being disposed of in ash ponds/pits or at the landfill posing a threat to human health and the environment. In this study, the physico-chemical properties of coal ash were analysed and based on that, their potential reuse were identified. The two methods implemented for this study were mainly the leaching method to reduce the level of heavy metals to the permissible limit and thermal treatment to reduce the amount of carbon to the required level. The basic laboratory test performed were moisture content, pH and loss on ignition. The leaching behaviour of phosphate was also analysed through variation of acid concentration and time. Also, a set-up of an ash filter bed on a lab-

---

* Corresponding Author Email: s.utchanah@uom.ac.mu.

scale basis was modelled for treatment of dye wastewater where parameters such as COD and colour of the untreated dye wastewater and treated dye wastewater were tested and further compared. From the different interpretations obtained, the potential reuse of ash was identified.

Keywords: coal fly ash, CFA, coal bottom ash, CBA, moisture content, MC, pH, loss on ignition, LOI, leaching behaviour, COD, colour

INTRODUCTION

Background

From the trend of the fast growing energy demand, the annual coal consumption is boosting up at a faster rate. Coal being considered as the most widely used primary fuel currently fuels 40% of the world's energy demand and represents an imperative source for energy supply for the next three decades (World Energy Council, 2016). The three topmost countries dominating on coal are mainly China, India and United-States which together represents more than 70% of the global coal use (IER, 2013). According to International Energy Outlook Report (2016), the amount of coal consumed globally would escalate from 153 quadrillion Btu in 2012 to 169 quadrillion Btu in 2020 and to 180 Btu in 2040 respectively representing an average increase rate of 0.6% on a yearly basis. Coal is mostly preferred for power generation due to its cheaper cost and high calorific value. However, the major issue associated with it, is the millions of tonnes of ash generated which are either being landfilled or disposed of in large ash ponds or ash pits. Subsequently, this is posing a threat to human health and environment due to the high level of heavy metals present within the ash.

Some of the environmental impacts and health issues are as follows (Basu et al., 2009):

- Large hectares of land are being occupied due to the disposal of huge quantity of ash either as dry ash or in ash pond and this may lead to land scarcity and even decrease in land value.
- Since dry fly ash is difficult to handle due to its fine nature, it may be carried away by winds resulting in corrosion of different structures.
- It disturbs the environment through its impact in the soil, in the air and even in water sources.
- Inhalation of fly ash for a long period of time may cause serious health impact such as silicosis, fibrosis of lungs, bronchitis and pneumonitis.

Hence it is imperative to find possible ways for re-using the ash formed for various applications for a more sustainable environment.

Coal - Material Characterization

Coal is a black colour and hard substance formed due to the decomposition of prehistoric vegetation after being exposed and buried to high heat and pressure (McLeod, 2017; Andrei, 2016). According to WCA (2016), the quality of coal is determined by the type of vegetation the coal is originated from, its depth of burial, the temperature and pressure at that depth and the time for the formation of coal. Coal can be classified into four main categories (McLeod, 2017):

1. *Lignite:* It is the youngest form of coal and accounts for 17% of the world's coal reserves.
2. *Sub-bituminous:* It burns more cleanly compared to the other types due to its low sulphur content and accounts for 30% of the world's coal reserves.
3. *Bituminous:* It is harder and darker compared to lignite and sub-bituminous coal. It accounts for 52% of the world's coal reserves.

4. *Anthracite:* It is the most mature type of coal compared to the others and contain the highest amount of carbon. It accounts for about 1% of the world's coal reserves.

The Combustion Process and Ash Formation

During the combustion process, some minerals from the coal melts and decomposes while the remaining is just slag. The fine mineral particles are thermally treated, slagged and are carried by the flue gases which are further collected in electrostatic precipitators. Most of the slagged fine particles are spherical and are subjected to thermal shock as soon as they leave the flame zone enabling them to have a highly glassy surface which may vary for each particles (Joshi & Lohtia, 1993). Particles that do not melt are therefore only treated thermally retaining the original shape of the specific mineral while some are converted into new minerals due to heating and cooling. Most of the fly ash particles are amorphous although certain crystalline compound (quartz, mullite, magnetite and calcium compounds) may be present which accounts of 15% by weight of the fly ash (Joshi & Lohtia, 1993). As far as concerned for the unburnt carbon content in ash, it can be determined through the loss on ignition test.

The two main by-products formed when combustion takes place are mainly fly ash (FA) and bottom ash (BA). FA is the finest ash particle which is entrained in the flue gas and collected either by electrostatic precipitators or baghouse compared to BA which is the coarser discrete or fused particles; heavy in weight and is settled down on the grates in the combustion chamber. After combustion, the amount of FA formed accounts for 80-90% of the byproducts while BA accounts for only 10-20% (Tharaniyil, 2013). FA mainly consists of very fine glass-like particles ranging between 0.01–100 mm in size (Singh et al., 2014). They are normally referred to as cenospheres due to the presence of hollow spherical particles which are light in weight and tends to float on water surfaces (Sideris et al., 2018). Cenospheres are normally made up of silica, iron and aluminium with size ranging from 1-500 μm and colour ranging from white to dark grey (Sideris et al., 2018). FA particles are normally very light with a low to medium bulk density comprising of a large surface area.

According to ASTM 618, there are mainly two classes of coal fly ash: class F and Class C where the class F type is obtained when burning anthracite or bituminous coal while class C refers to ashes obtained when burning lignite or sub-bituminous coal and contains less than 50% of SiO_2, Al_2O_3 and Fe_2O_3 (Chun et al., 2007).

Variation in the chemical composition of FA and BA from different power plants can be observed based on the following factors (Dina et al., 1988; Sideris et al., 2018):

- Quality and chemical composition of the coal
- Degree of coal pulverization
- Types of furnace and conditions under which coal is burnt
- The way of handling and storage of fly ash

Physical and Chemical Properties of Fly Ash

Physical Properties

The different physical properties are specific gravity, moisture content, grain size distribution, free swell index, colour, and water holding capacity as well as plasticity and shrinkage limit.

Specific Gravity

Specific gravity is an imperative physical property required for the use of ashes in various applications. It is usually related to the shape, colour and chemical composition of the ash (Sideris et al., 2018). As per Upadhyay & Kamal (2007), specific gravity ranges from a low value of

1.90 for a sub-bituminous ash to a high value of 2.96 for an iron-rich bituminous ash. A low specific gravity could either be the consequence of large numbers of hollow cenospheres present from which the entrapped air cannot be removed or a variation in the chemical composition, in particular the iron content; or both (Emilliani & Dygku, 2007). In terms of composition, fly ash has a greater amount of specific gravity as compared to bottom ash.

Moisture Content

Moisture content is one of the important physical properties when dealing with the ash cementitious properties. Normally fly ash especially class C type hardens due to hydration of its cementitious compound when there is presence of moisture in the ash (Sideris et al., 2018). Therefore dry ash must be used when mixed with cement. According to ASTM C618, to prevent caking of ash during storage, a moisture content limited to 3% is specified (Sideris et al., 2018).

Grain Size Distribution

Grain size distribution helps in indicating whether a material is well graded, poorly graded, fine or coarse (Sasak, 2006). As such, it is a method to classify the ashes. Constituting of some sand-size fraction, coal ashes are predominantly silt sized. They can be categorized as sandy silt to silty sand depending on their grain size distribution (Upadhyay & Kamal, 2007). The silt size normally ranges between 1 and 100 μm and normally enhance the fluidity and workability of fresh concrete (Sideris et al., 2018). Fly ash normally ranges from 0.03 mm to 0.001 mm and is classified as mostly fine silt to fine sand sizes while bottom ash ranges from 10 mm and 0.075 mm and is classified as fine gravel to fine sand sizes (Awang et al., 2012).

Free Swell Index

Free swell index is used to identify swelling soils. Considering coal ashes, due to a very low clay-size fraction, the free swell index is considered as negligible (Upadhyay & Kamal, 2007).

Colour

Depending on their chemical and mineral constituents, fly ashes can range from tan to dark grey in colour. Tan and light colours are due to the high concentration of lime while the brownish colour is mainly due to the iron concentration (Joshi & Lohtia, 1993; Sideris et al., 2018). If there is an augmentation in the unburned content, the colour can fall in the range of dark grey to black.

Water Holding Capacity (WHC)

WHC plays an important role when considering sandy soil. Due to high permeability in sandy soil, there are water and nutrients losses. By addition of fly ash to such soil, the permeability reduces and the water and nutrients retentivity is improved (Dhindsa et al., 2015). This is beneficial to the plant and thus increases the yield.

Plasticity Limit and Shrinkage Limit

Plasticity, shrinkage limit and liquid limit are important properties for the Atterberg limit. The shrinkage limit is dependent on moisture content i.e., the soil will start to swell if the shrinkage limit is exceeded and increase in moisture content beyond plastic limit will cause the soil to form a plastic mass resulting in additional swelling (Global Road Technology,

2017). The soil will remain in the plastic state until its liquid limit is exceeded causing it to flow as a viscous liquid (Global Road Technology, 2017). In the construction sector, proper control of the Atterberg limit especially on expansive soil is important to prevent damage of structures as with increase in swell index, the soil strength decreases.

According to Global Road Technology (2017), plasticity for expansive soil is obtained through the plastic index (PI) which exhibit different plasticity at different moisture content and this can be illustrated in Table 1 which shows the level of plasticity in different types of soil.

Table 1. Plasticity index of different types of soil

Soil type	PI %	Type of plastic soil
Clay loam	10-20	Medium plastic soil
Silty Clay	20-35	Highly plastic soil
Mainly clay	>35	Very high plastic soil

Table 2. Physical properties of coal fly ash (CFA)

Parameters	Fly Ash
Bulk Density (gm/cc)	0.9-1.3
Specific Gravity	1.6-2.6
Plasticity	Lower or non-plastic
Shrinkage Limit	Higher
Grain size	Major fine sand/silt and small per cent of clay size particles
Clay (per cent)	Negligible
Free Swell Index	Very low
Classification (Texture)	Sandy silt to silty loam
Water Holding Capacity (WHC) (per cent)	40-60
Porosity (per cent)	30-65
Surface Area (m^2/kg)	500-5000
Lime reactivity (MPa)	1-8

Table 3. Normal range of chemical composition for fly ash produced from different coal types (expressed as percent by weight)

Component	Bituminous	Sub-bituminous	Lignite
SiO_2	20-60	40-60	15-45
Al_2O_3	5-35	20-30	10-25
Fe_2O_3	10-40	4-10	4-15
CaO	1-12	5-30	15-40
MgO	0-5	1-6	3-10
SO_3	0-4	0-2	0-10
Na_2O	0-4	0-2	0-6
K_2O	0-3	0-4	0-4
LOI	0-15	0-3	0-5

The physico-chemical properties of CFA are illustrated in Table 2 and 3 respectively (Data Source: Bombatkar et al., 2017; FHWA, 2016; Tharaniyil, 2013).

The Different Chemical Constituents Present in Fly Ash and Their Associated Impacts upon Application

The aim of this study is to investigate the physical and chemical properties of coal fly ash (CFA) and coal bottom ash (CBA) as well as find possible ways to implement its potential reuse as a useful by product. The specific objectives are as follows:

The setting up of an ash filter bed on a lab scale basis to treat dye wastewater where parameters such as colour and COD for the treated and untreated samples are tested and further compared, basic laboratory test such as moisture content (MC), pH, loss on ignition (LOI) were conducted. A leaching analysis was also performed to analyse the amount of phosphate leached with varying concentration of sulphuric acid and increasing time.

Table 4. Analysis of the different chemical constituents present in fly ash

Components	Impacts
Na_2O & K_2O	• Fly ashes containing large quantities of Na_2O & K_2O are likely to be very reactive due to the fact that the alkalis increase the pH of the pore solution when they are amalgamated in concrete and the high pH speeds up the dissolution of the glass in the fly ash (Thomas, 2007). • A high rate of alkalis in the fly ash may result into the expansion and damage of hardened concrete (Jarvis, 2015) which can also lead to crack formations. • Normally used as a soil stabilizer due to its high alkalinity to neutralize the acidic condition of the soil and increase crop yield.
MgO	• If an excess of magnesia (MgO) is present in cementitious materials, it may give rise to unsoundness, that is, an unwanted volume of change when utilized in concrete (Thomas, 2007). • However, its alkaline properties make it a good soil stabilizer as pointed out for Na_2O & K_2O.
Al_2O_3	• Cements comprising of large amounts of alumina have a very high initial strength which make them resistant to sulfate attacks as well as weak acids. However, they can lose a considerable part of their strength and durability with time, hence, causing damage to numerous structures (Chatterjee, 2018). • But, their appropriate ratio with silica depicts the level of pozzolanic activity. • Li (2008) pointed out that Al_2O_3 in ashes are vital to produce coagulants in the treatment of wastewater.
SiO_2	• Use as filler to enhance the interface transition zone and to produce denser concrete (Zareei et al., 2017). • Addition of silica can be categorized by production of high strength concretes, exothermic rate reduction, more corrosion resistance, increase sustained strength of concrete permeated with chloride ions and sulfates in the range of 2–4 times, less permeability, more durability, and less interaction between alkali cement with aggregates (Zareei et al., 2017).

Components	Impacts
SO_3	• In general, the sulfate content in fly ash falls between a range limit of less than 0.1% to 5% SO_3. • According to ASTM C618, if fly ash is used in concrete, its sulfate content needs to be limited to 5% SO_3 (Thomas, 2007).
CaO	• Lignite or sub-bituminous coals produce fly ashes with high calcium content. The calcium content of the fly ash is perhaps the best indicator of how the fly ash will behave in concrete. • Fly ash hardens when mixed with water due to the hydration reaction taking place. FA with high calcium content can be used solely as a cementing material to produce moderate strength concrete (Thomas, 2007). • The amount of calcium present in fly ash is a key determinant of the latter's effectiveness to decrease the heat of hydration, control expansions caused by alkali-silica reactions and resist sulfate attacks (Thomas, 2007). • Thomas (2007) also highlighted that if the amount of CaO in cementitious materials is more than required when used in concrete, it may give rise to an undesirable volume change.
Fe_2O_3	• A small amount of the iron found in fly ash affects the pozzolanic activity and may lead to stains at high concentrations. • As per Ramezanianpour (2014), the pozzolan properties will be further improved if Fe_2O_3 is separated from silica and aluminuim. • Fe_2O_3 makes concrete sulfate resistant. • The presence Fe_2O_3 in ashes plays a key role in producing coagulants to treat waste water (Li, 2008).

Materials and Methods

Origin of CFA and CBA

Around 5 kg of CFA and CBA were collected from Alteo co-generation plant situated at Beau-Champs, Flacq, Mauritius.

The different analysis performed were MC, pH, LOI, leaching behaviour of phosphate and set-up of a CFA filter bed for treatment of dye wastewater.

Moisture Content (MC)

Following the procedure of ASTM D2974, 50 g of ash sample was weighed and placed in trays. The trays were then placed in an oven at 105^0C for 24 hrs. The trays were then removed and reweighed and the different values obtained were noted. This test was repeated thrice and an average for the different weight was then calculated. The moisture content was then calculated by using the formula below:

$$\boldsymbol{MC}\ (\%) = \frac{\boldsymbol{A - B}}{\boldsymbol{A}} \times \mathbf{100}$$

where:

A: Total wet weight of ash (without tray), g
B: Total dry weight of ash (without tray), g

pH

Following the Donahoe et al., (2007) method, 5 g of ash was added to distilled water in a ratio of 1: 3 (i.e., to 10 g of ash, 30 g of distilled water was added) and was left to rest for 24 hrs. The pH was then found using a pH meter by dipping the probe in the sample and allowing the pH to stabilize.

LOI

5 g of dry ash was weighed on a 4 sf electronic balance in a silica crucible (pre-weighed) and placed in a furnace. This procedure was

repeated for 6 different crucibles which were placed in the furnace together at 550^{0}C for 90 min. At an interval of 15 min, one crucible was removed, allowed to cool for 2 min and then weighed. This experiment was repeated at varying temperature (550^{0}C, 650^{0}C, 750^{0}C, 850^{0}C and 950^{0}C) and removed at varying time (15, 30, 45, 60, 75 and 90 min). At varying time, the values were noted. The use of different crucibles at varying time was to prevent losses while weighing. This procedure was taken from ASTM D7348 but was slightly modified by varying the temperature and time intervals.

Leaching Behaviour of Phosphate

- The phosphate concentration was found by following the published method given by Huang et al. (2011) with slight modification. 5g of ash devoid of moisture was weighed and placed in a conical flask. 50 ml of distilled water was then added to the content using a measuring cylinder followed by 50 ml of 1M H_2SO_4 and was allowed to shake on a mechanical shaker at 150 rpm for 30 min.
- This procedure was repeated with 4 different conical flask at different time interval (30 min, 1hr, 2hr, and 3hr) to analyze the concentration of phosphate (PO_4^{3-}) leached with increase in time. After mechanical shaking, the suspension was filtered and the clear filtrate was collected for the analysis of PO_4^{3-}.
- The above procedure was then repeated with 100 ml of 1M H_2SO_4.
- The test was also performed without addition of H_2SO_4 acid which was taken as a control where only 100 ml distilled water was used to leach the ash following the above procedure in step i and ii.
- From the different test performed and filtrate obtained, the concentration of phosphate was tested as per the HACH method.

Coal Ash Filter Bed for Dye Wastewater Treatment

Theory behind Dye Wastewater Treatment Using Ash as Filter Bed

The model as shown in Figure 1 was set up based on the method used by Sanas and Gawande (2016) for treatment of wastewater using fly ash. For this set-up, coal fly ash was used for treating dye wastewater which has the ability to reduce the colour, the chemical oxygen demand (COD) and also the total suspended solid (TSS).

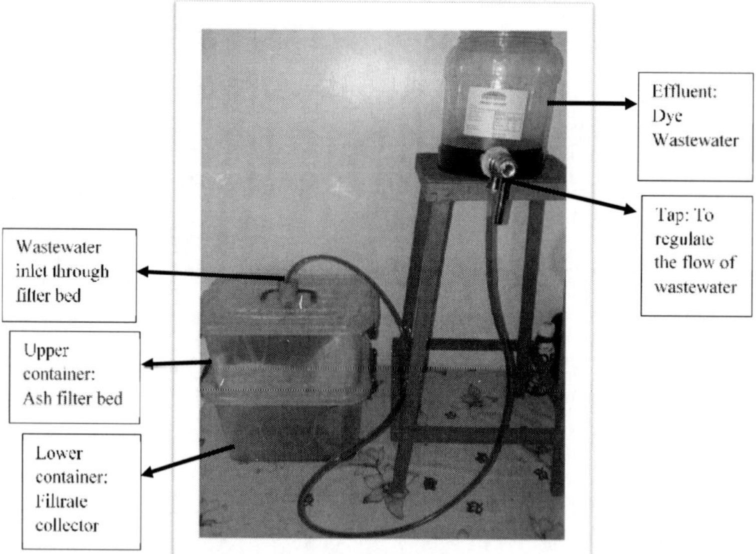

Figure 1. Ash filter bed for dye wastewater treatment.

Firstly, the ash was washed several times with distilled water to remove adhering organic material and the plant substance. It was then allowed to dry at room temperature for 24 hrs. After drying, the ash was used for the filtering process where it was set as filter bed of 5 cm and 10 cm thick respectively. This was obtained by filling a storage container (plastic container) having small holes at the bottom surface at a height of 5 cm. A fabric material was placed at the bottom surface of the container before placing the ash so as to prevent dispersion of ash through the pores.

Another plastic container was placed just below the storage container containing ash so as to collect the filtrate passing through the ash. The above container acts the inlet where wastewater is introduced and percolates through the ash while the container below acts as the outlet where the treated filtrate is obtained. This set up was repeated with an ash bed height of 10 cm. Ash acts as a good adsorbent for wastewater treatment due to its high carbon content.

Effluent Collection

The dye wastewater was obtained from world knits ltd situated at Phoenix, Mauritus, where 5 L of the wastewater sample was collected by the operators at different time interval in a day.

Procedure

- The ash was first washed with distilled water and left to dry for 24 hrs at room temperature.
- The upper container was filled with ash up to 5 cm.
- Then around 1 L of the dye wastewater sample was allowed to percolate through the ash filter bed.
- The treated effluent obtained was tested for COD and colour.
- Step i to iv was repeated but this time the height of the ash filter bed was changed to 10 cm.
- Step i-iv was repeated with CBA. The different results obtained were further compared.

RESULTS AND DISCUSSION

Moisture Content

For the moisture content test, the amount of water in the sample was determined which was expressed in terms of percentage of the wet weight of the sample. From this test, a moisture content of 26.1% was obtained for CFA and 27.3% for CBA.

MC of ashes is considered as a critical parameter as different MC have different impacts on different applications. Considering soil stabilization, MC plays an important role especially on expansive soil as this type of soil tends to undergo excessive swelling and shrinkage with increase in MC and this represents a potential danger mainly for civil structures (Hasan et al., 2016; Wubshet & Tadesse, 2014). Based on literature, the optimum moisture content (OMC) of soil and bagasse normally lies between 40-50% (Wubshet & Tadesse, 2014).

Ashes with high MC can be added to the soil together with lime so as not to affect the strength of the soil as lime normally helps in the treatment of expansive soil by reducing soil expansion, increasing strength, decreasing plasticity index, swell and shrinkage potential and also control volume changes (Wubshet & Tadesse, 2014) where these parameters are referred to as the Atterberg limit.

Considering the application of wet ash for construction purposes, mainly concrete, from literature it was found that high moisture content tends to harden cement because of hydration of its cementitious compound (Sideris et al., 2018). Moreover, in fly ash, due to the high carbon content, when in wet state has a tendency to segregate the carbon particle which may float on top of the cement slurry and causes it to thicken. Hence more water has to be added to allow it to flow (Global Cement, 2012). For this reason dry ash is mostly preferred in cement. For ashes with high moisture content, they should be thermally treated before applied to cement so as not to affect the cement strength. Also for storage purposes, if ashes with high moisture content has to be stored for a long period of time, then care should be taken to prevent caking. Normally a moisture content limited to 3% is preferred for storage.

pH

After allowing the sample to stand in distilled water for 24 hrs, it was observed that an alkaline pH of 9.21 was obtained for CFA and 8.32 for CBA. Thus fly ash with such alkalinity nature, can be used for soil treatment by raising its pH and thereby neutralize the acidic soil. Fly ash normally acts as liming agent and provides the plant with the required nutrients. According to IPCC, fly ash is mostly preferred as liming agent to neutralize the soil compared to agricultural lime as application of lime contributes to global warming when applied to the soil due to the carbon content in it which is released as carbon-dioxide to the atmosphere (Kishor et al., 2010). Thus FA is mostly preferred as it reduces carbon-dioxide emissions and thus reduces global warming.

LOI

The LOI test is performed to determine the percentage of unburnt carbon in the ash sample. For this test, the temperature of furnace was varied from 550^0C-950^0C and at each temperature, the time was also varied whereby after every 15 min, one crucible was removed and weighed. The different data obtained at different time interval and temperatures for CFA and CBA can be illustrated in Figure 2 and Figure 3 respectively.

For LOI based on the test conducted, it can be observed that the percentage LOI increases with higher temperature. Moreover, the amount of carbon burnt is faster in FA compared to BA. This might be because of their physical properties basically as the particle size of BA is normally coarser compared to FA which are finer, hence has a tendency to burn faster.

Unburnt carbon in ashes normally play an important role when used in cement. From literature, a 5-6% LOI is mostly preferred when used in cement as increasing amount of unburnt carbon has many adverse effect of the cement paste. The water requirement and rheological properties of concrete are normally affected with increase in unburnt carbon in FA as higher the amount of carbon, the more water is required to obtain a paste with a normal consistency (Mohebbi et al., 2015). Moreover, increase in unburnt carbon tends to absorb air entraining admixtures (AEAs) increasing the water requirement again.

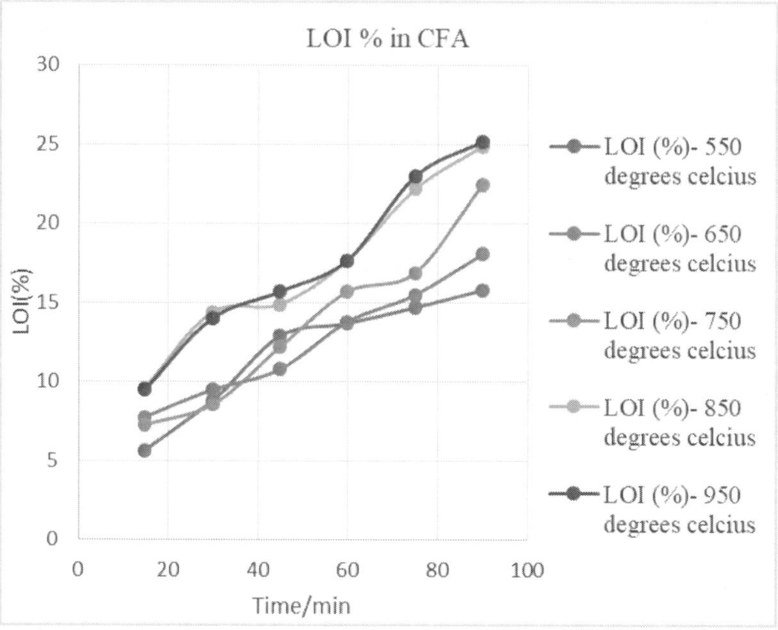

Figure 2. LOI in CBA with variation in time.

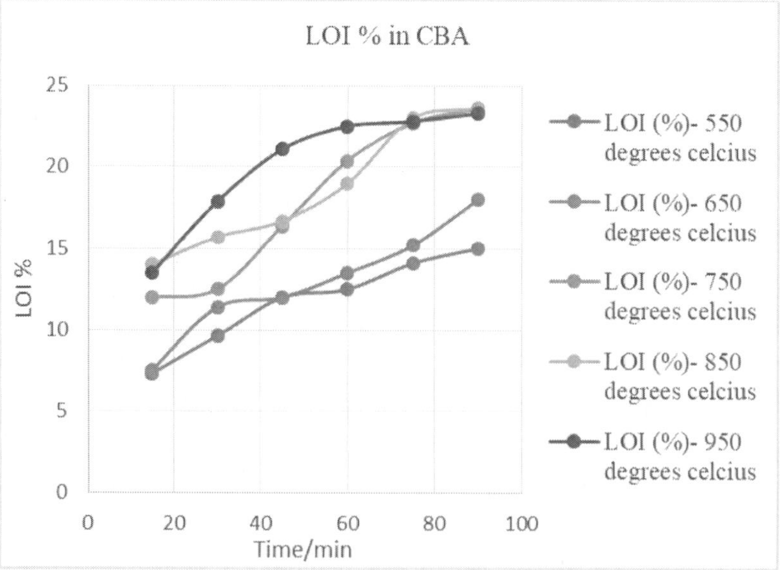

Figure 3. LOI in CFA with variation in time.

They also tend to change the colour of concrete to black or dark grey as the carbon particles tend to float to the top of the concrete finishing process. Hence before being used as ad-mixture in cement, a carbon-burn out plant can be introduced for thermal treatment of the ash so that the amount of unburnt carbon is reduced to the required limit and can hence be used as cement additives.

High amount of unburnt carbon are mostly beneficial for wastewater treatment as it has a tendency to adsorb more toxic metals, COD and other impurities. Higher the amount of carbon, better will be the adsorption of impurities.

Leaching Behaviour of Phosphate with Variation in Time and Concentration of Acid

In this experiment, the behaviour of nitrate and phosphate leached with variation in time and concentration of acid was investigated and this can be illustrated in Figure 4.

Based on the different test conducted and as illustrated on Figure 4, it can be observed that the concentration of phosphate in CFA and CBA leached increases with time. The higher the shaking time, the higher is the concentration and this behaviour can be explained by the fact that, with increasing time, the diffusion rate is enhanced and hence more phosphate is being extracted thereby increasing the leaching efficiency.

The pH is another important parameter in this experiment as it can be observed that higher the volume of H_2SO_4 added to the sample, better is the extraction rate. From the test conducted, sample to acid ratio of 5:100 (g/ml) gives higher concentration of phosphate (PO_4^{3-}) compared to a ratio of 5:50 (g/ml). When these results are compared to that of water, that is ash sample to water ratio of 5:100 (g/ml), it can be observed that the reaction is slower for CFA and that for CBA is almost negligible causing the concentration of PO_4^{3-} to be comparatively lower from that obtained by addition of acid. Thus, it can be deduced that diffusion rate is enhanced at lower pH, hence increasing the leaching rate.

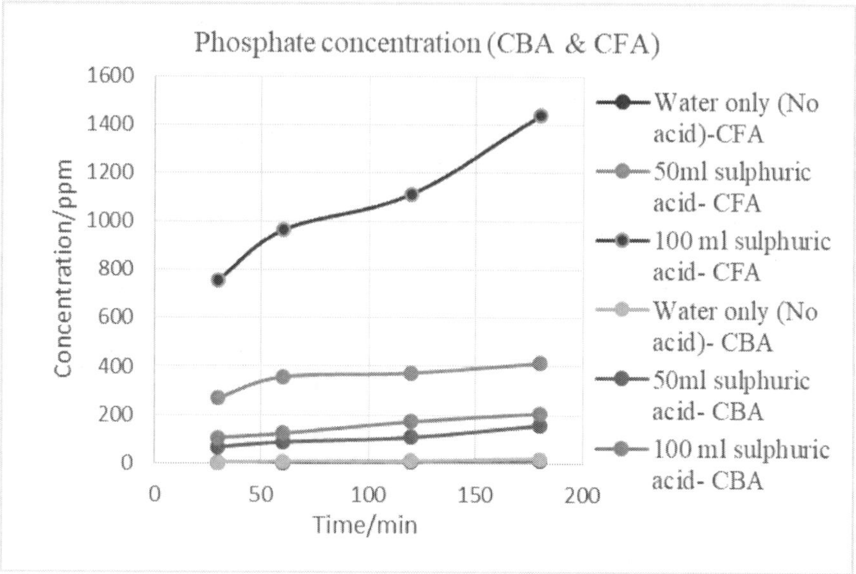

Figure 4. Concentration of phosphate leached from CFA and CBA at varying time interval.

From this experiment, it can be observed that the leaching efficiency is dependent on both time and concentration of acid as a sample to acid ratio of 5: 100 (g/ml) and a shaking time of 3 hrs has proven to have better extraction properties of phosphate whereby the highest phosphate concentration was obtained.

The concentration of phosphate is very high for coal ash which is beyond the permissible limits as per the environment protection act for effluent discharge for surface water. Hence if ashes are disposed of near surface water courses and leached, this can lead to eutrophication causing damage to the ecosystem. It is better to use them for agricultural purposes as their leachability through water is very low and hence in case of heavy rainfall, it won't affect the land or underground water. Thus it is more suitable to be used as fertilizer. Phosphate are also beneficial in concrete as they prevent acidic erosion of concrete.

COD and Colour

Dye wastewater (WW) was treated by passing it through different height of ash filter bed and the treated sample obtained was then tested for COD and colour removal. The data obtained are illustrated below and analyzed.

From this test, it was observed that fly ash is very efficient for reducing the COD and colour level of wastewater. The colour of the wastewater changes from dark blue to almost colourless after being passed through fly ash. And for the COD level, CFA is more efficient compared to CBA and this may be due to the physical properties of the ash. Fly ash is of very fine size and the distance between the particles are almost negligible as compared to bottom ash which are larger in size with larger distances in between the particles. Hence the absorption capacity of fly ash tends to increase due to less void spaces between the particles which help to absorb more heavy metals and impurities, reducing the colour and COD to a high extent. Also the amount of unburnt carbon in FA is more compared to BA (from LOI test). Higher the amount of carbon, better is the absorption efficiency impurities in the wastewater.

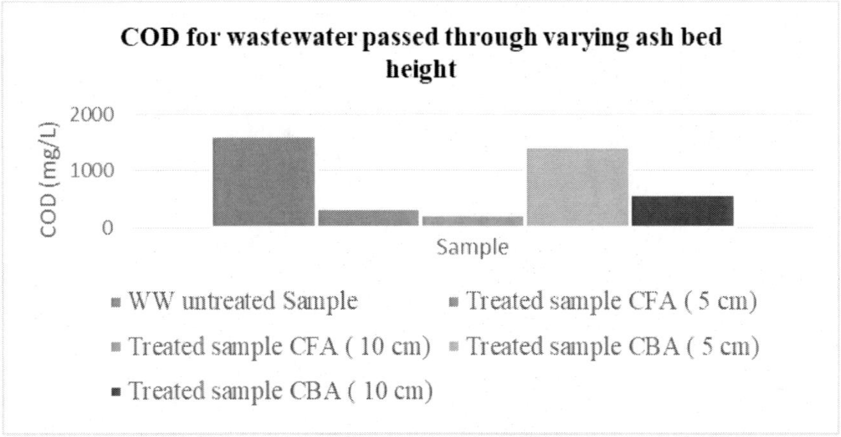

Figure 5. Variation of COD for treated and untreated sample.

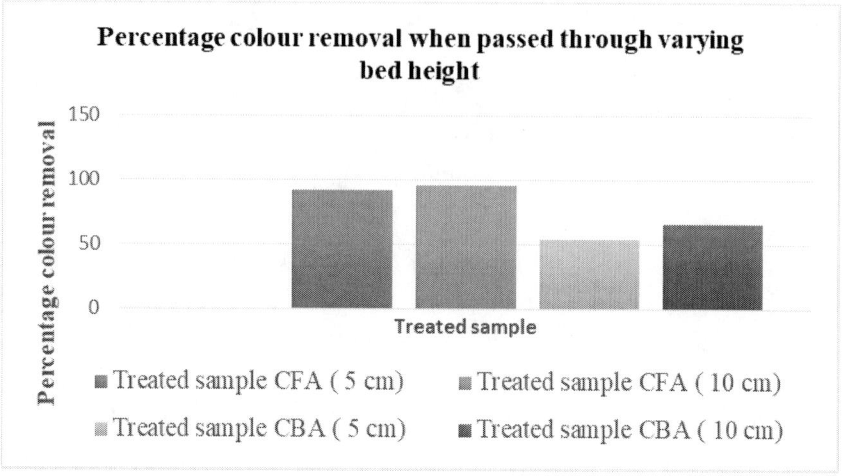

Figure 6. Percentage colour removal from wastewater through varying ash bed height.

By varying the bed height, it was observed that with an ash bed height of 10 cm, the COD level reduced significantly as compared to that of the 5 cm height. For the 10 cm height for CFA, the percentage of COD removed was 86.6% as compared to the 5 cm height which was 80.2%. Even when using CBA, the COD level reduces above 60% with a bed height of 10 cm.

Considering the test for colour removal its was observed that when passing the dye wastewater through an CFA bed of 10 cm, the percentage colour removed is above 90% and when using CBA, a percentage colour removal of above 60% is obtained. Thus fly ash has proven to be more efficient for treating dye wastewater as maximum colour can be removed and even the COD level is decreased significantly.

Conclusion

The main aim of this study was to analyse the physico-chemical properties of coal ash and identify possible ways for their potential re-use as a useful by-product. The two test method employed were the leaching method for the removal of heavy metals and hence reducing the toxicity level in ash to meet permissible limit and secondly, thermal treatment to reduce the carbon content in ash so that it can be used in construction sector most specifically as additive in cement.

Based on the different interpretations, the most suitable way for ash re-use is for soil stabilization due to its alkaline pH where acidic soil can be treated. Moreover it can be used in the construction sector whereby the amount of carbon has to be reduced to the required level through thermal treatment and can further be used as additive in cement and as far as concern for bottom ash, they can be used for structural fills mainly for roads or base of buildings due to their coarser size. Also due to the high concentration of phosphorus in ashes, the appropriate proportion can be added to fertilizer and used for agricultural purposes. Furthermore, the high amount of carbon makes it a potential adsorbent for removal of impurities and toxic metals found in wastewater. Therefore for a more sustainable environment, these techniques can be applied whereby our landfill is preserved and the problem of land pollution is tackled.

REFERENCES

Andrei, Mihai. 2016. *How coal is formed.* 15 January. Accessed September 10, 2017. https://www.zmescience.com/science/geology/how-coal-is-formed/.

Awang, A. R., Marto, A. & Makhtar, A. M., 2012. Morphological and Strength Properties of Tanjung Bin. *International Journal on Advance Science Engineering Information Technology,* 2(2).

Basu, Manisha, Pande, Manish, Bhadoria, P. B. S. and Mahapatra, S. C. 2009. "Potential fly-ash utilization in agriculture: A global review." *ScienceDirect* 1173-1186.

Bombatkar, A. G., Kautikrao, S. and Bhosale, A. R. 2017. "Stabilization of Soil by using Fly Ash." *Imperial Journal of Interdisciplinary Research.*

Chatterjee, D. 2018. *High Alumina Cement.* Accessed January 20, 2018. https://www.scribd.com/doc/155463169/High-Alumina-Cement.

Chun, Y. M., Claisse, P., Naik, T. R. & Ganjian, E. 2007. *Sustainable Construction Materials and Technologies.* London: Taylor & Francis Group.

Dhindsa, H. S., Sharma, R. D. & Kumar, R. 2016. Role of fly ash in improving soil physical properties and yield of wheat (Triticum aestivum). *Agricultural Research Communication Centre,* 36(2), pp. 97-101.

Dina, E. M., Donald, J. L. & Leonard, H. W. 1988. A review of physical, chemical, and biological properties of fly ash and effects on agricultural ecosystems. *ScienceDirect*, Volume 74, pp. 1-37.

Donahoe, R. J., Bhattacharya, S., Patel, D. and Ladwig, K. J. 2007. *Chemical Fixation of Trace Elements in Coal Fly Ash.* 7-10 May. Accessed September 8, 2017. http://www.flyash.info/2007/161donahoe.pdf.

Emilliani, A. G. & Dygku, S. I. 2010. Geotechnical Properties of Fly Ash and its Application on. *UNIMAS E-Journal of Civil Engineering,* 1(2).

FHWA. 2016. *User Guidelines for Waste and Byproduct Materials in Pavement Construction: Coal Fly Ash.* 3 August. Accessed October 2017, 15. https://www.fhwa.dot.gov/publications/research/infrastructure/pavements/97148/016.cfm.

Global cement. 2012. *Power plant ashes as substitute raw materials.* 19 November. Accessed March 5, 2018. http://www.globalcement.com/magazine/articles/735-power-plant-ashes-as-substitute-raw-materials.

Global Road Technology. 2017. *Atterberg Limits.* Accessed March 15, 2018. https://www.globalroadtechnology.com/atterberg-limits/.

Hasan, H., Dang, L., Khabbaz, H., Fatahi, B. and Terzaghi, S. 2016. "Remediation of Expansive Soils Using Agricultural Waste Bagasse Ash." *ScienceDirect*: Elsevier 1368-1375.

Huang, K. et al., 2011. Leaching behavior of heavy metals with hydrochloric acid from fly ash generated in municipal waste incineration plants. *ScienceDirect*: Elsevier, 21(6), pp. 1422-1427.

IER. 2013. *Latest Analysis: China Will Continue to use Coal as a Major Generating Source.* 6 August. Accessed June 15, 2018.

https://www.instituteforenergyresearch.org/analysis/china-will-continue-to-use-coal-as-a-major-generating-source/.

Jarvis, J. B. 2015. *The Effect of Low-Level Sodium Concentrations on the Use of Fly Ash as a cement Replacement in Concrete.* 5 May. Accessed December 8, 2017. http://www.flyash.info/2015/221-jarvis-2015.pdf.

Joshi, R. C. & Lohita, R. P., 1993. *Fly Ash in Concrete: Production, Properties and Uses.* Alberta, Canada: Gordon and Breach Science Publishers.

Justnes, K. H., Soutsos, M. and Sui, T. 2018. *Fly Ash.* Accessed March 4, 2018.www.springer.com/cda/content/document/cda.../9783319706054-c2.pdf?SGWID.

Kishor, P., Ghosh, A. K. and Kumar, D. 2010. "Use of Flyash in Agriculture: A Way to Improve Soil Fertility and its Productivity." *Asian Journal of Agricultural Research* 1-14.

Li, L. et al., 2008. Production of a new wastewater treatment coagulant from fly ash with concomitant flue gas scrubbing, *Journal of Hazardous Materials 162 (2009) 1430–1437,* Ames, United States: Elsevier.

McLeod, Charlotte. 2017. *The 4 Types of Coal and Their Uses.* 8 March. Accessed September 10, 2017. https://investingnews.com/daily/resource-investing/industrial-metalsinvesting/coal-investing/types-of-coal/.

Mohebbi, M., Rajabipour, F. & Scheetz, B. E. 2015. Reliability of Loss on Ignition (LOI) test For Determining the Unburned Carbon Content in Fly Ash, *World Coal Ash Conference (WOCA)* Nashville (2015).

Ramezanianpour, A. A. 2014. *Cement Replacement Materials,* Berlin: Springer Geochemistry/Mineralogy.

Sanas, M. V. & Gawande, S. M. 2016. Fly Ash using in Waste Water Treatment. *International Journal of Emerging Engineering Research and Technology,* 4(6), pp. 11-14.

Sasak, T. K. 2006. *Physical and Geotechnical Properties of Coal Ash.* Accessed March 4, 2018. http://docplayer.net/49811737-Faculty-of-engineering.html.

Singh, R. P. et al. 2014. Biological Responses of Agricultural Soils. *Reviews of Environmental Contamination and Toxicology,* Volume 232.

Tharaniyil, R. 2013. *Coal Combustion Products Utilization Handbook.* 3rd ed. United States of America: We Energies Publication.

Thomas, M. 2007. *Optimizing the Use of Fly Ash in Concrete.* Accessed January 20, 2018.http://www.cement.org/docs/default-source/fc_concrete_technology/is548-optimizing-the-use-of-fly-ash-concrete.pdf.

Upadhyay, A. and Kamal, M. 2007. *Characterization and Utilization of Fly Ash.* Accessed November 22, 2017. http://ethesis.nitrkl.ac.in/4293/1/Characterization_and_utilization_of_fly_ash.pdf.

WCA. 2016. *World Coal Association, What is coal?* Accessed October 4, 2017. https://www.worldcoal.org/coal/what-coal.

WEC. 2016. *World Energy Resource 2016.* 3 october. Accessed September 12, 2017. https://www.worldenergy.org/wp-content/uploads/2016/10/World-Energy-Resources-Full-report-2016.10.03.pdf.

Wubshet, M. and Tadesse, S. 2014. "Stabilization of Expansive Soil Using Bagasse Ash & Lime." *EEA.*

Zareei, S. A., Ameri, F., Dorostkar, F. & Ahmadi, M. 2017. Rice husk ash as a partial replacement of cement in high strength concrete containing micro silica: Evaluating durability and mechanical properties. *ScienceDirect*: Elsevier, Volume 7, pp. 73-81.

In: Coal Fly Ash
Editor: Miriam Gray

ISBN: 978-1-53614-511-3

Chapter 2

INVESTIGATION OF THE USE OF COAL FLY ASH AS RAW MATERIAL IN CEMENT MORTARS

M. D. La Rubia, J. Cabrera-Díaz, R. Órpez and D. Eliche-Quesada*
Department of Chemical, Environmental and Materials Engineering, Higher Polytechnic School of Jaén, University of Jaen, Jaén, Spain

ABSTRACT

Recycling is a strategy to consume the waste of several industries as raw material. Coal-fired power plants produce millions of tons of fly ash that cause environmental problems. In addition, these waste materials need large deposit areas with high cost.

In this study, the synergy between Portland cement and coal fly ash was investigated. Raw materials were characterized in terms of chemical and mineralogical composition and physical properties. Mortars were manufactured and the effect of the addition of different ratios of coal fly ash (0-80 wt %) as a partial substitute of Portland cement was studied. The mineralogical composition of the sustainable mortars has been analyzed by XRD and FT-IR. Physical and mechanical properties as bulk density, water absorption by capillarity, flexural and compressive strength of hardener mortars have been determined according to EN standards. The results show that the bulk density of the cement mortars is slightly affected by the coal fly ash addition but leads to an increase of the water absorption by capillarity after of 7 and 28 of curing time. The densification of fly coal ash mortar during the curing age contributed to a significant reduction of the porosity and the water absorption by capillarity. The flexural and compressive strength decrease as the ratio of coal fly ash increases with respect to the control mortar, except for the mortar that incorporates 20 wt % of coal fly ash. The mortars containing up to 20 wt % of coal fly ash offer an opportunity to recycle wastes and present physical and mechanical properties similar to control mortars, obtaining economical and environmental benefits.

Keywords: coal fly ash, cement mortars, recycling, physical and mechanical properties

1. Introduction

There is growing international concern over the use of alternative components as raw material in the production of construction materials. One of the main objectives of the construction industry is to seek innovative products and develop new materials with no environmental pollution implications and high rates of energy efficiency. In this sense, the use of secondary raw materials is being favored by Governments and legislation and rules are being improved to facilitate this use [1].

The use of coal combustion products in Europe is being influenced by new regulations and environmental legislation, since the Waste Framework Directive, 2008/98/EC defined what product of coal combustion is a by-product and when a material loses the condition of residue [2].

Fly ash is a residue from the combustion of coal in thermal power plants during the generation of electricity. This residue is obtained by electrostatic or mechanical precipitation of dust like particles from the flue gas. Fly ash can be considered a cheap raw material since it is a residue of an industrial activity that is carried out in significant amounts. Its main cost is transportation, so its use is clearly indicated in areas near the production centers [3].

The production of coal fly ashes in the European Union (EU 15) in 2016 was 25.741 million tons. Actually the utilization of coal fly ashes in the construction industry is around 43% of the production. They are widely used in the building and construction industry (Table 1) [4]. In the majority of cases the coal fly ashes are used as a replacement for cement in concrete and as an aggregate in road construction. Additional applications are as structural filler, soil amendment and as fertilizers.

All these applications offer several environmental benefits because the energy demand is reduced, the need to quarry or mine primary resources is avoid and the emissions to atmosphere are decreased, for example CO_2, which result from the manufacturing process of the products which are replaced.

Table 1. Uses of ash coal fly in the construction industry [4]

Uses	**kt**	%
Cement raw material	1881	16.56
Blended cement	1936	17.04
Concrete addition	4638	40.82
Aerated concreted blocks	433	3.81
Non-aerated concreted blocks	195	1.72
Bricks +ceramics	7	0.06
Grouting	229	2.02
Asphalt filler	22	0.19
Subgrade stabilisation	71	0.62
Pavement base course	11	0.10
General engineering fill	12	0.11
Structural fill	569	5.01
Soil amendment	43	0.38
Infill	167	1.47
Other uses	187	1.65

Fly ash can chemically react with $Ca(OH)_2$ (CH in cement chemist notation), in a process known as pozzolanic reaction, to produce the secondary C-S-H (Calcium-Silicate-Hydrates), which are the components that provide mechanical resistance. This pozzolanic activity makes possible to incorporate or replace cement with ashes in the formulation of construction materials with the associated environmental benefits and costs [5, 6]. At present, the use of fly ash as an admixture in concrete is regulated by the European standard EN 450-1 (2005) [7].

There are studies with concretes with high content of fly ash (between 40 and 80 of the total weight of binder material, cement and ashes). The addition of ash allows the reduction of the amount of cement in the concrete, decreases the heat so the temperature gradients are smaller and the danger of superficial fissuring is reduced [8, 9]. The resistance of the concrete with fly ash is lower compared to that of no addition at early ages, but over time, the resistance is approaching, even reaching the long-term resistance of the concrete with fly ash [10].

The replacement of cement by fly ash in the manufacture of mortars has also been studied. However, the addition of fly ash are not higher than 40 wt %. Kuroda et al. (2000) showed that the incorporation of fly ash affected the microstructures of the interface between the aggregates and the cementitious matrix, especially the porosity and pore structures. Tosti et al. [11], evaluated the mechanical and environmental properties of cement mortars containing fly ash (20 wt % and 40 wt %) from biomass combustion as a secondary cementitious material. The substitution of 20 wt % of cement complied with the reference standard for compressive strength and the mortars maintained good technical and environmental performance with a 40 wt % of reduction of the carbon footprint. Chen et al. [6], developed mortars with pulverized fly ash and sewage sludge ash as suitable cement with a maximum amount of replacement of 20 wt %. In several cases the use of coal fly ash bring extra quality and higher performance and it adds a green label to construction due to the natural resources preservation [12].

The objective of this work is to study the synergistic effects that occur between high percentages of coal fly ash and cement when used in the manufacture of sustainable materials. For this purpose, mortars with different ratios of coal fly ash (0-80 wt %) as cement-like material were manufactured. The mortars obtained have been characterized in order to determinate the possible fields of applications.

2. Materials and Methods

2.1. Characterization of the Raw Materials

Physical and chemical characterization of the raw materials is described in this section. The raw materials used in this investigation were Portland cement (PC) type CEM II/A-L 42.5 R supplied by Holcim (Spain), coal fly ash (CFA) from Carboneras Power Plant (Almería, Spain) as partial replacement of cement and natural sand CEN DIN-EN-196-1 from Instituto Torroja (Madrid, Spain).

The chemical composition of the raw materials was determined by X-ray fluorescence (XRF) by a spectrometer Philips Magix Pro (PW-2440). Crystalline phases were evaluated in a X-ray diffractometry with an X'Pert Pro MPD automated diffractometer (PANanalytical) equipped with Ge (111) primary monochromator, using monochromatic Cu Kα radiation (λ = 1.5406 Å) and an X'Celerator detector. The content of carbonates was determined by calcimetry. Blaine method was used to determine the specific area of the raw materials [13]. The relative density value was obtained by the Le Chatelier`s volumetric method. The pH was determined by a solid pH-meter PCE-PH20S, (PCE Instrument).

2.2. Pastes and Mortars Preparation and Testing

Pastes were prepared in this study with different portions of cement and coal fly ash. Different percentages of cement were replaced by coal fly ash (0, 20, 40 60 and 80 wt %) and the ratio water/(cement+coal fly ash) was kept at 0.5 in all the mixes. The pastes were designed as xCP-yCFA where x indicated the percentage of Portland cement and y the percentage of coal fly ash. A summary of the mix proportions is provided in Table 2.

The pastes were cured during 7 and 28 days, then they were analyzed by DRX and Fourier transform infrared spectroscopy (FT-IR) to study the formation of hydrated compounds of cement paste such as calcium silicate hydrate (C-S-H), calcium hydroxide ($Ca(OH)_2$), ettringite and also the existence of carbonation. For that purpose, a spectrometer Bruker Tensor 27 with a spectral range of 370 – 4500 cm^{-1}, was used.

Mortars were prepared according to EN standards [14] with natural sand CEN DIN-EN-196-1. The mortars were made with Portland cement type CEM II/A-L 42.5 R and the water/ (cement+coal ash fly) relation (W/C) was 0.5 by weight. The cement was partially replaced by coal ash fly and five different percentages of substitution were used: 0, 20, 40, 60 and 80 wt %. The mixing procedure used was defined by the standard EN 1015-3 (1999) [15]. Mortar mixture proportions are reported in Table 3. Two series of six prismatic specimens were prepared for each mortar mixture, 40 x 40 x 160 mm size, by casting them in steel forms. They were cured in water at 20ºC up to 7 and 28 days and dried before testing.

Table 2. Composition of the pastes mixes

Pastes	Cement (g)	Coal fly ash (g)	Water (g)
100CP-0CAF	40	0	20
80CP-20CAF	32	8	20
60CP-40CAF	24	16	20
40CP-60CAF	16	24	20
20CP-80CAF	8	32	20

Table 3. Composition of the mortars

Mortars	Cement (g)	Coal Fly Ash (g)	Water (g)	Sand (g)
100CP-0CAF	450	0	225	1350
80CP-20CAF	360	90	225	1350
60CP-40CAF	270	180	225	1350
40CP-60CAF	180	270	225	1350
20CP-80CAF	90	360	225	1350

Physical and mechanical properties after 7 and 28 days of curing were determined. Apparent density determination, were carried out according to the standard EN 1015-10 (2000) [16], capillary water absorption tests according to EN 1015-18 (2003) [17]. The mechanical properties were determined using the Universal Testing Machine MTS 810 according to EN 1015-11 (2007) [14]. The mortars were designed as xCP-yCAF-z, where x indicated the percentage of cement and y the percentage of coal ash fly and z the curing time 7 or 28 days.

3. Results and Discussion

3.1. Characterization of Raw Materials Portland Cement (PC) and Coal Fly Ash (CFA)

Table 4 shows the results of pH, specific surface, relative density, carbonate content and pH of raw materials, Portland cement type II (CEM II) and coal fly ash.

Both PC and CFA have high pH values, being the PC values higher. High pH values favor the pozzolanic reaction between silica and alumina present in the raw materials in the wet phase. The carbonate content is higher in Portland cement than in the coal fly ash. The presence of carbonates in Portland cement is due to the addition of limestone.

Table 4. pH, specific surface, relative density, carbonate content and pH of raw materials, Portland cement type II (CEM II) and coal fly ash

Raw material	pH	Specific Surface (cm2 g-1)	Carbonate content (%)	Relative density (kg cm-3)
CEM II	11.8	3670	16.98	2840
CFA	9.3	4273	11.98 ± 0.58	2480

The specific surface values of the raw materials obtained by the Blaine method [13] are higher for the coal fly ash (4273 cm^2/g) than for the Portland cement (3670 cm^2/g). High specific surfaces favor the reaction between the aluminosilicates present in the pozzolan and the calcium hydroxide generated in the hydration of the Portland cement, resulting in the formation of a greater amount of hydrated calcium silicates (CSH) responsible for the mechanical resistance and the impermeability of the mortars.

Both raw materials have similar values of relative density, being coal fly ash slightly lighter than Portland cement.

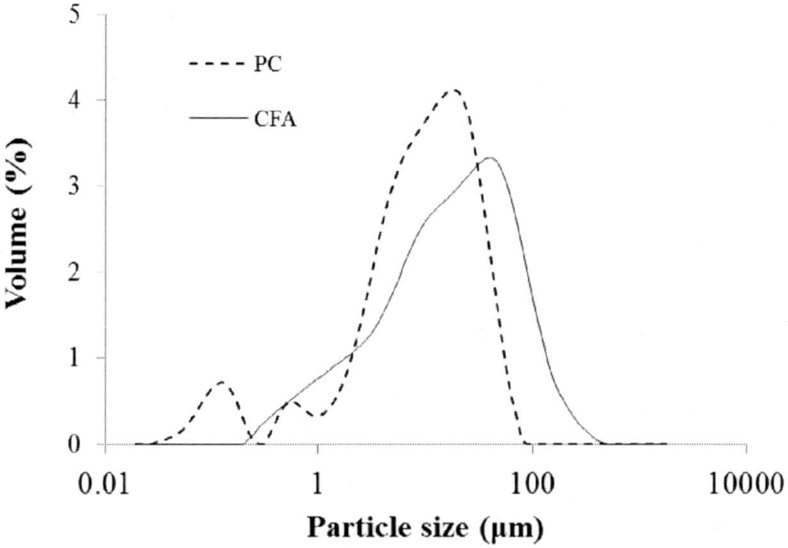

Figure 1. Particle size distribution of raw materials, Portland cement (PC) and coal fly ash (CFA).

Table 5. Particle size distribution in raw materials

Particle size distribution (mm)	PC (% part.)	CFA (% part.)
Clay content (< 0.002)	11.95	9.56
Silt content (0.002 – 0.063)	86.30	69.62
Sand content (0.063 - 2)	1.5	20.82

Table 6. Chemical composition of the raw materials determined by XRF

Oxide content (%)	PC (%)	CFA (%)
SiO2	21.330	53.340
Al2O3	5.890	21.020
Fe2O3	3.870	7.200
CaO	58.320	3.770
MgO	1.390	1.570
K2O	0.880	1.870
Na2O	0.910	-
MnO	0.060	0.050
TiO2	0.430	0.900
P2O5	0.080	-
P2O6	-	0.400
SO3	4.030	0.820
SrO	-	-
Cu	-	0.010
Cl	0.080	-
Sr	0.070	0.140
Cr	0.020	0 – 2
Ni	0.010	0.020
Cu	0.030	0.010
Zn	0.030	0.020
Pb	-	0.010
Rb	-	0.010
Ba	0.070	0.210
Zr (ppm)	82.700	-
LOI	2.610	4.740

The particle size distribution of the raw materials is shown in Figure 1. Portland cement and coal fly ash have similar particle size distribution, with an average particle size D_{50} of 10.6 μm for the PC and 19.6 μm for the CFA. The main particle size fraction in both raw materials is silt-size (0.002 - 0.063 mm) (Table 5). The particle size of the raw materials is a crucial factor affecting the pozzolanic activity; a decrease in the particle size implies an increase of the specific surface of the material and then a rise of the hydraulic reaction rate [18].

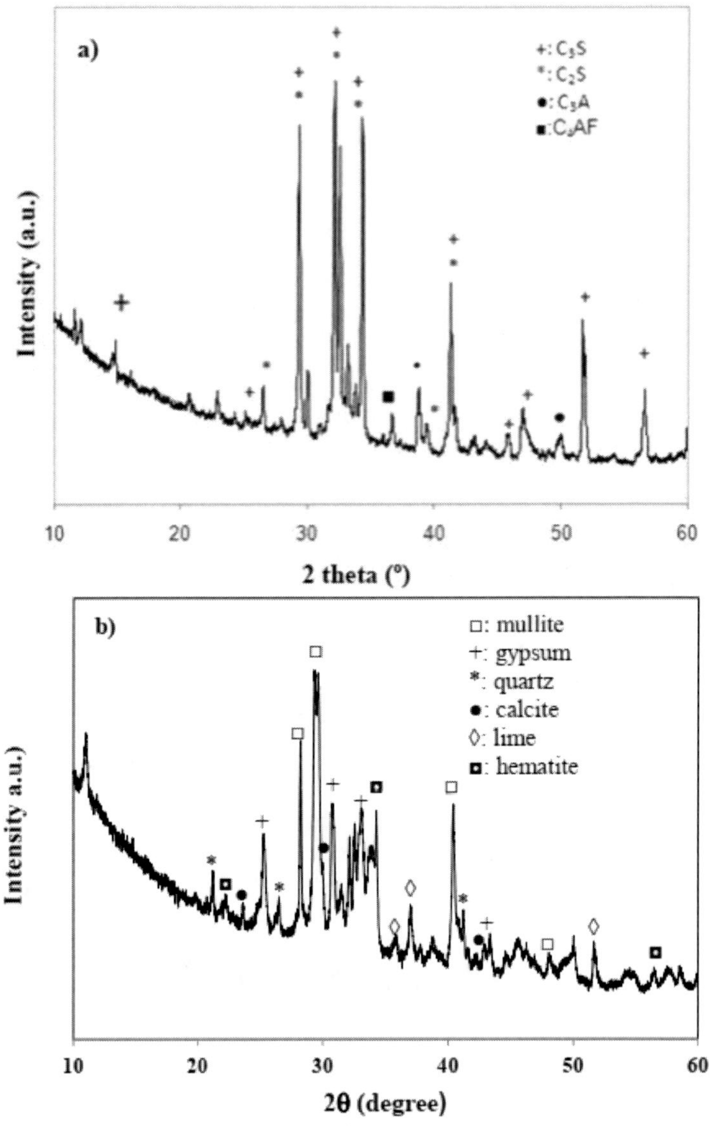

Figure 2. DRX of (a) Portland cement (PC) and (b) coal fly ash (CFA).

The chemical composition of the raw materials determined by FRX (Table 6) indicates that Portland cement has a high content of CaO (58.8%) and SiO_2 (21.4%), with a lower proportion of Al_2O_3 (5.9%), Fe_2O_3 (3.9%), and sulfate SO_3 (4.0%). The coal fly ashes are composed mainly of SiO_2

(53.3%) and Al_2O_3 (21.0%). The content of Fe_2O_3, MgO, K_2O and CaO in decreasing order varies between 1% and 8%, with low sulfate content. The CFA can be classified according to their composition as class F ash, alumina silica of pozzolanic character.

In the diffraction pattern of the Portland cement (Figure 2a), the characteristic diffraction peaks of tricalcium silicate (C_3S) and dicalcium silicate (C_2S), known as alita and belite, respectively, can be observed. The components tetracalcic aluminoferrite, 4 CaO Al_2O_3 Fe_2O_3 (C_4AF), and the tricalcium aluminate, 3CaO Al_2O_3 (C_3A), are found in smaller quantity. The coal fly ash is composed of partially vitreous mineral particles due to the high temperature reached during the combustion process and, the subsequent fast cooling. Therefore, the crystalline phases are in lower proportions, as the DRX pattern shows (Figure 2b). The crystalline components in CFA are quartz, hematite, mullite and alumina silicate. The diffraction peaks corresponding to calcite, calcium sulfate and calcium oxide are also observed in smaller quantity.

3.2. Microstructural Characterization of the Pastes

The diffractograms of the control paste with only Portland cement (Figure 3) shows the diffraction peaks corresponding to the hydration products of the PC, such as, ettringite and calcium hydroxide. All these peaks are observed after 7 and 28 days of curing time. In addition, the diffraction peaks due to the presence of the bicalcium silicate (C_2S), tricalcium silicate (C_3S) from unhydrated cement and calcite, are also observed. The most intense peak corresponds to the calcium hydroxide, which is highly crystalline. However, the formation of hydrated calcium silicate gel is not appreciated, by the XRD technique, due to its non-crystalline nature. For that reason, its formation is analyzed by the infrared spectroscopy (FT-IR) analysis technique.

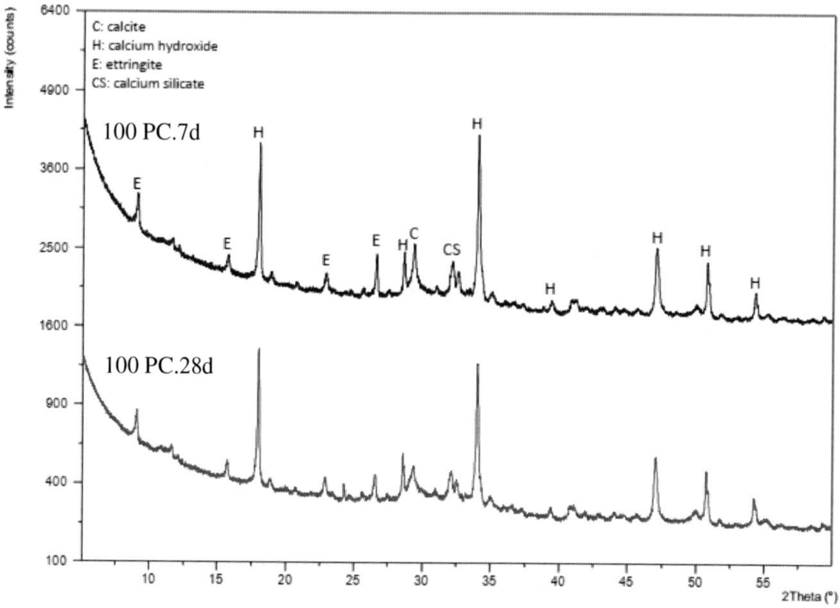

Figure 3. DRX of 100PC pastes after 7 and 28 curing days.

The analysis of the diffraction peaks also shows the evolution of the hydration reaction because the intensity of the peaks is modified with the curing time. As the curing time increases from 7 to 28 days, a higher intensity of the diffraction peaks corresponding to calcium hydroxide is observed, also a slight decrease of the intensity of the peaks of the calcium silicates is observed indicating a greater progression of the hydration reaction of the Portland cement. In addition, a slight increase of the intensity of the diffraction peak corresponding to $CaCO_3$ is observed, probably due to the fact that part of the calcium in the calcium silicates have reacted with the atmospheric CO_2 in the air and the superficial carbonation of the pastes has increased with the curing time.

The substitution of 20, 40, 60 and 80 wt % of Portland cement by coal fly ash (Figure 4), modifies the crystalline phases present in the cement pastes that incorporate ash as pozzolans, as well as, the intensity of the diffraction peaks. In the PC-CFA pastes, not only the diffraction peaks corresponding to the hydration products of the PC, ettringite and calcium hydroxide are observed but also the diffraction peaks corresponding to the unreacted calcium silicates and calcium carbonate. In addition, the CHS gel, due to its amorphous nature is not observed, as in the control mortar occurred. It can be observed, that as the content of coal fly ash increases, the intensity of the diffraction peaks corresponding to calcium hydroxide decreases. This means that the calcium hydroxide produced in the hydration reaction of the Portland cement reacts with the aluminosilicates present in the pozzolana in the presence of water, producing cementitious components, calcium silicates hydrated. In addition, when the quantities of ash incorporated increase, the diffraction peaks corresponding to the non-reactive crystalline phases appear, such as quartz.

Regarding the influence of the curing time (Figure 5), small differences are observed; the calcium silicates decrease slightly due to the progress of the pozzolanic reaction and the calcium hydroxide diminishes due to its reaction with the acidic compounds of the CFA.

Figure 6 shows the FT-IR spectra of the control paste containing only Portland cement at 7 and 28 days of curing. The most important peaks identified are: gypsum, calcium hydroxide, calcium carbonate, hydrated calcium silicate and ettringite. In the region of 3800-3000 cm^{-1}, a peak between approximately 3644-3638 cm^{-1} is observed, and a very broad band centered at approximately 3445-3402 cm^{-1} related to the elongation vibration of OH. The first peak is assigned to the free OH^- in the calcium hydroxide ($Ca(OH)_2$), product of the hydration reaction of the calcium silicates. The broad band in this region can be assigned to the elongation vibrations produced by the water molecules in the CSH (hydrated calcium silicate gel), obtained by the hydration of Portland cement.

The second region, between 1800 and 1200 cm^{-1}, has a small broad band centered at 1635 cm^{-1} and another intense band centered at 1413 cm^{-1}. The band centered at 1635 cm^{-1} is assigned to the vibration modes of the water molecules and the band centered at 1413 cm^{-1} to the asymmetric and symmetric vibration modes of $(CO_3)^{2-}$. This confirms the presence of carbonates. This band is more intense after 28 days of curing, according to the XRD diffractograms.

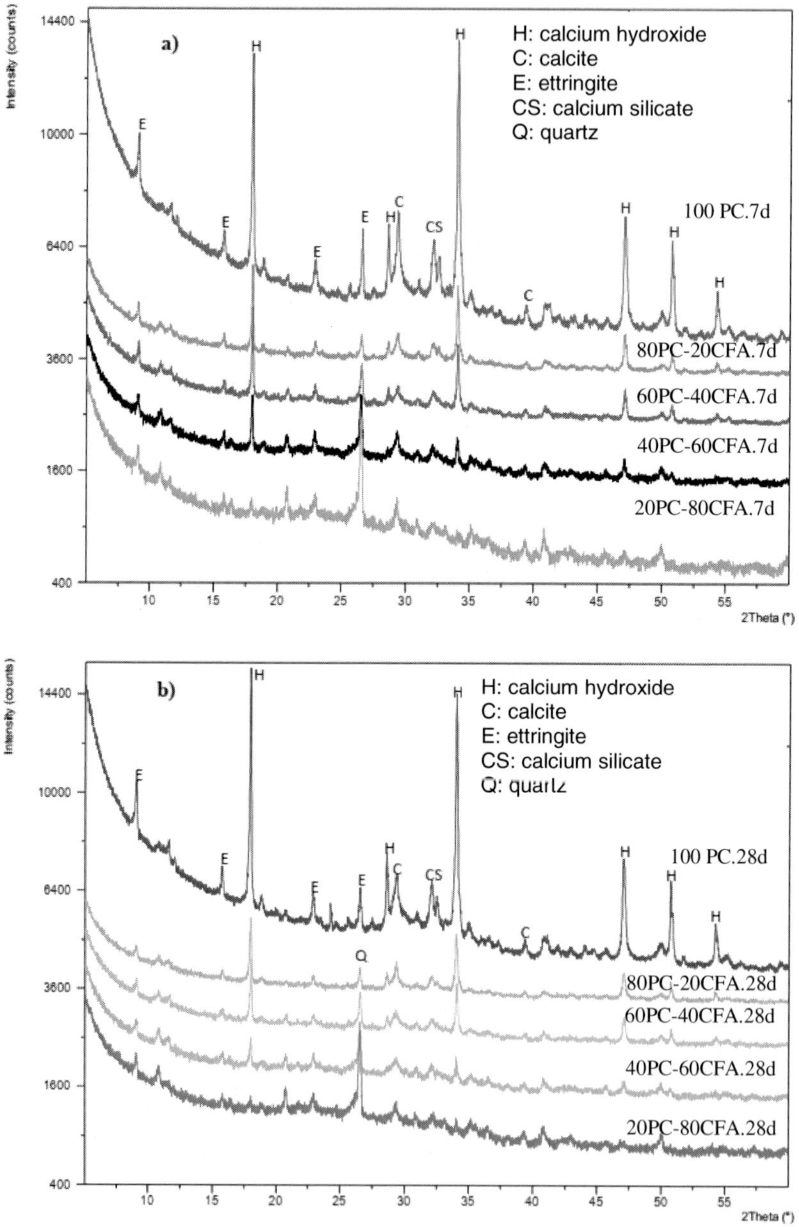

Figure 4. DRX of (a) xPC-yCFA pastes after 7 curing days and (b) xPC-yCFA pastes after 28 curing days.

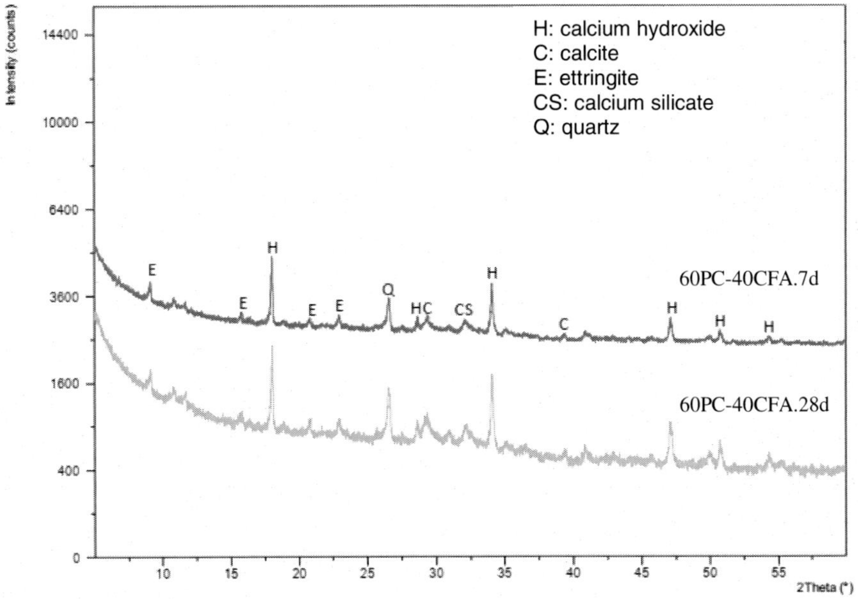

Figure 5. DRX of 60PC-40CFA pastes after 7 and 28 curing days.

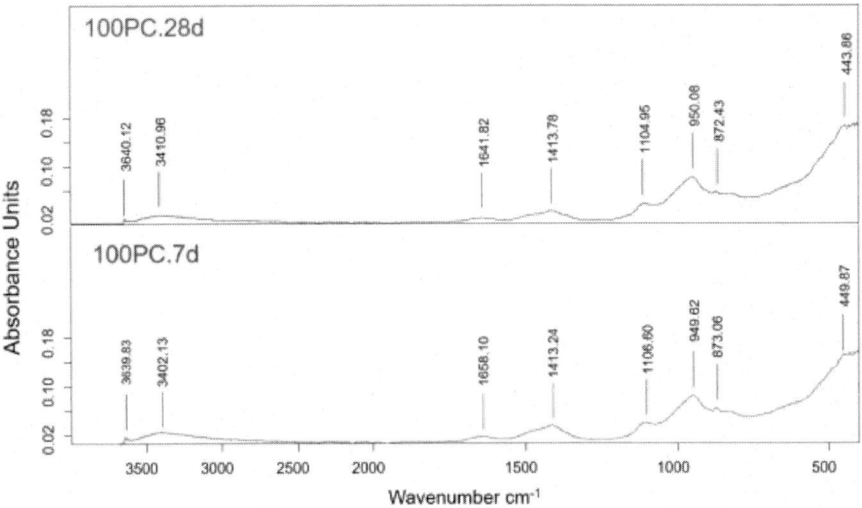

Figure 6. FTIR of 100PC pastes after 7 and 28 curing days.

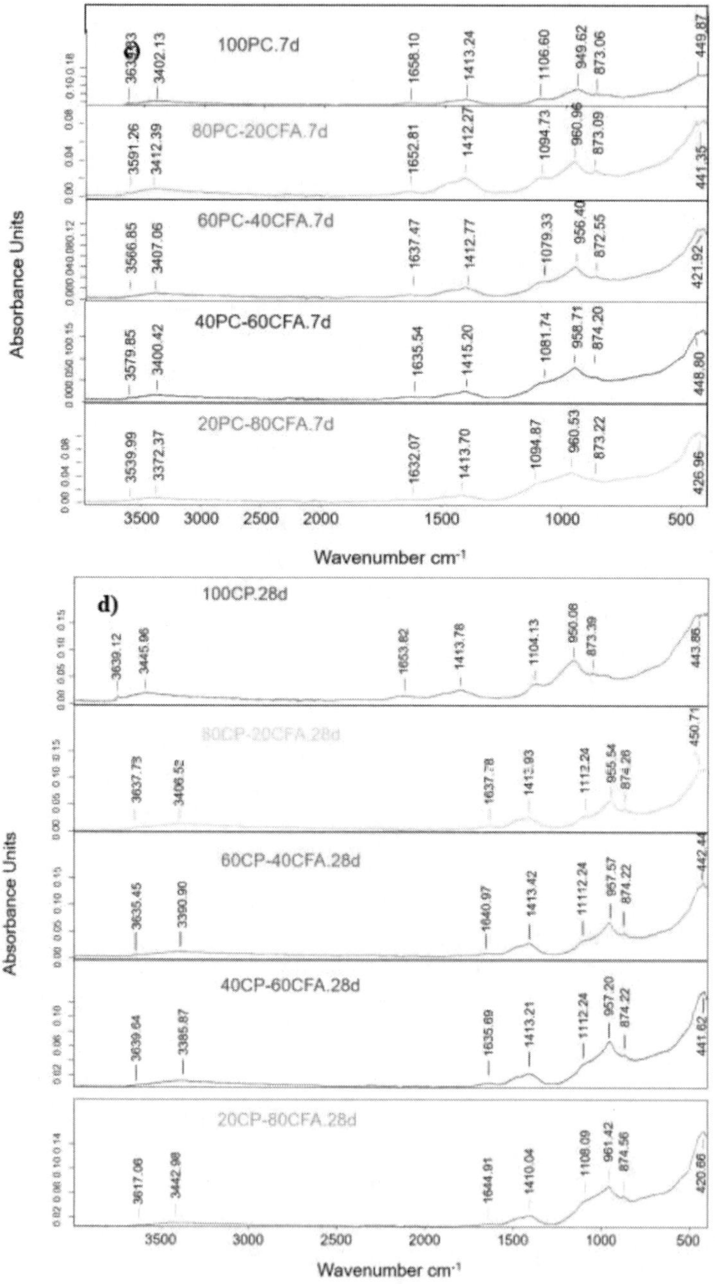

Figura 7. FTIR of (a) xPC-yCFA pastes after 7 curing days and (b) xPC-yCFA pastes after 28 curing days.

The third region, between 1200 and 800 cm^{-1}, has a wide band centered at 1110-1103 cm^{-1} which is characteristic of the Si—O stretching modes indicating the formation of the CSH gel. This band is more intense with longer curing time, which indicates the progress of the hydration of the paste over time. The band corresponding to the sulfates is also observed centered at approximately 1110 cm^{-1} due to the vibration of the SO_4^{2-} group of the sulfates. The bands at approximately 876 cm^{-1} are assigned to the secondary modes of bending of the carbonates. The band centered between 460 and 430 cm^{-1} can be due to the vibrations produced by the Si—O and Al—O stretches of the aluminate and silicate groups [19].

The FT-IR spectra of the pastes with the CFA additions (Figures 7) are barely modified with respect to the FT-IR spectra of the control paste. The main differences are the decrease and near disappearance of the centered peak at approximately 3620 cm^{-1} assigned to the elongation vibration modes of free OH^- that is present in calcium hydroxide ($Ca(OH)_2$). This fact is due to its combination with the aluminosilicates present in the ashes to form CHS gel. The broadband centered at approximately 3400 cm^{-1} assigned to the vibrations of elongations produced by the water molecules in the CSH gel and, the hydration reaction of the pozzolans with the calcium hydroxide becomes more pronounced. Although the band centered at approximately 1640 cm^{-1} assigned to the modes of bending of water molecules is hardly modified. The characteristic bands of the modes of vibration of the $(CO_3)^{2-}$ centered at approximately 1410 cm^{-1} and 870 cm^{-1} decrease as the content of CFA increases in the pastes, indicating a lower carbonation of these pastes. The band centered at approximately 1110 cm^{-1} due to the vibration of the SO_4^{2-} group of the sulfates is hardly modified. However a slight increase of the band centered to approximately 960 cm^{-1} characteristic of the ways of stretching of the Si—O from both the gel CHS and from the ashes is observed, being more intense in the pastes with higher ash content. The band centered at approximately 440 cm^{-1}, which is attributed to the vibrations produced by the Si—O and Al—O stretches of the aluminate and silicate groups, increases slightly with the ash amounts.

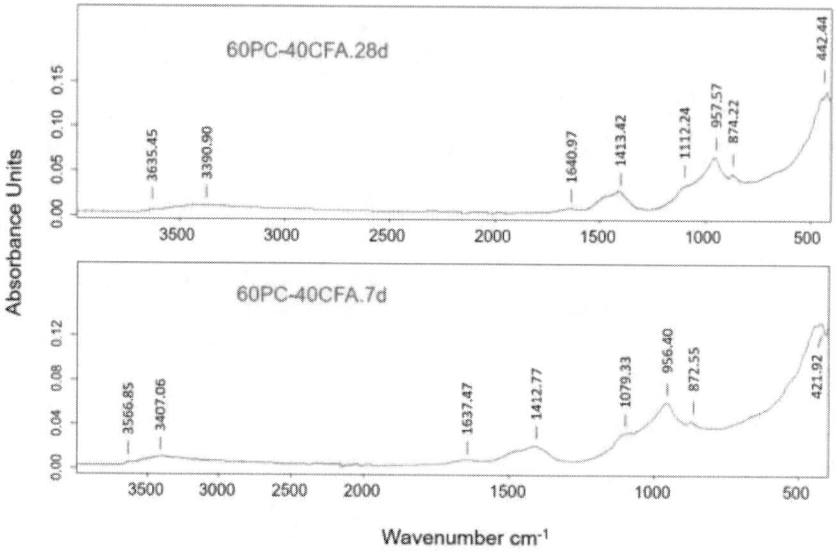

Figure 8. FTIR of 60PC-40CFA pastes after 7 and 28 curing days.

Regarding the influence of the curing time (Figure 8), a small increase of the characteristic bands of the CHS gel can be observed. This indicate an advance of the hydration reaction of the pastes.

3.3. Characterization of the Mortars

The bulk density of the mortars can be observed in Table 7. The bulk density of the control mortars with only Portland cement (PC) are 2216 kg/m^3 and 2280 kg/m^3 after 7 and 28 days of curing, respectively. The bulk density of the mortars with coal fly ash does not change significantly with respect to the control mortar. Only a reduction of 3.0 and 5.2% with the addition of 80 wt % of CFA after 7 and 28 days of curing was observed. The slight decrease in the bulk density of the mortars that incorporated CFA as cementitious material can be due to the lower relative density of the raw material CFA (2480 kg/m^3) with respect to Portland cement (2840 Kg/m^3).

Table 7. Bulk density and water absorption by capillarity of sustainable mortars after 7 and 28 curing days

Mortars	Curing time		Curing time	
	7 days	28 days	7 days	28 days
	Bulk density (kg cm-3)		Water absorption by capillarity (Kg m-2 min-0.5)	
100PC	2217 ± 3.99	2280 ± 8.85	0.05 ± 0.01	0.03 ± 0.01
80PC-20CFA	2290 ± 20.97	2270 ± 1.45	0.07 ± 0.03	0.04 ± 0.01
60PC-40CFA	2232 ± 3.97	2247 ± 8.51	0.07 ± 0.02	0.03 ± 0.01
40PC-60CFA	2214 ± 3.67	2199 ± 20.88	0.10 ± 0.02	0.09 ± 0.01
20PC-80CFA	2148 ± 4.45	2160 ± 1.79	0.33 ± 0.01	0.35 ± 0.02

Regarding to the influence of the curing time, the bulk density of the control mortars and the mortars with the addition of coal fly ash for a curing time of 7 days are slightly lower than those obtained after 28 days of curing time. The common reason for this is than at short curing times, 7 days, the hydration reactions is uncompleted and the quantity of hydrated calcium silicates, CSH phases, is low and therefore, the network of pores is not compacted.

The capacity of the mortars to absorb water is closed linked to the morphology (size, interconnection) and the amount of pores that the mortars' structure. The values of the coefficient of water absorption by capillarity of the control mortars are 0.052 and 0.03 $kg/m^2min^{0.5}$, for curing times of 7 and 28 days, respectively (Table 7). The addition of CFA produces a slight increase in the coefficient of water absorption by capillarity increasing from 0.07 and 0.4 $kg/m^2min^{0.5}$ with the addition of 20 wt % of CFA and curing times of 7 and 28 days respectively to 0.33 and 0.35 $kg/m^2min^{0.5}$ with the incorporation of 80 wt % of CFA for curing times of 7 and 28 days respectively (Table 7). It can be seen how the incorporation of up to 60 wt % of CFA increases slightly this physical property, while the addition of 80 wt % of residue produces a very remarkable increase of the values of the coefficient of water absorption by capillarity.

Table 8. Masonry rendering and plastering mortars classification according to their water absorption by capillarity

Category	Water absorption by capillarity(Kg m-2 min-0.5)
W0	Not specified
W1	C ≤ 0,4
W2	C ≤ 0,2

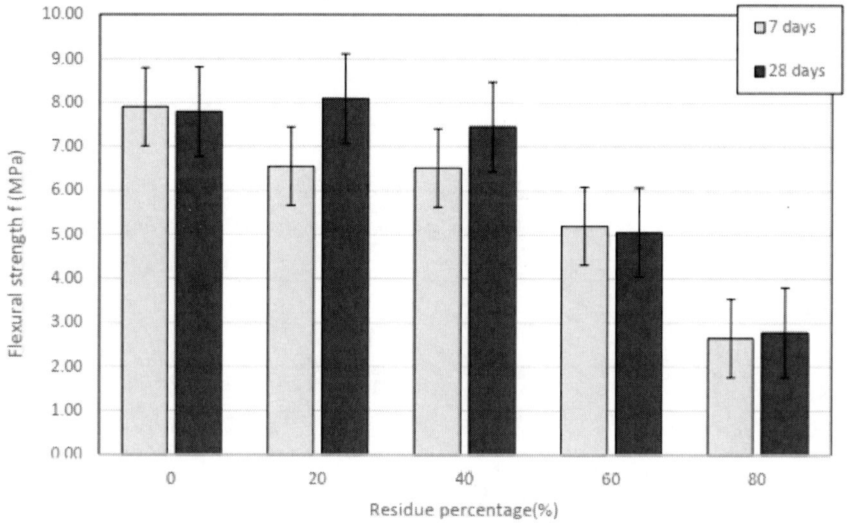

Figure 9. Flexual strength of the xPC-yCFA mortars for 7 and 28 curing days.

Regarding the influence of the curing time, it is observed that the coefficient of water absorption by capillarity with the incorporation of CFA as a cementitious material is higher after 7 days of curing time. This fact could indicate, according to the bulk density data, that after 7 days of curing the hydrated calcium silicates, responsible of the compacted pore network formation is still being produced, as the XRD data showed.

The EN 998-1: 2010 standard [20] classifies masonry rendering and plastering mortars in three categories according to the water absorption coefficient due to capillarity action (Table 8). All mortar specimens that incorporate CFA after 7 and 28 days of curing, except the mortars 20PC-80CFA, have a water absorption coefficient smaller than 0.20 $kg/m^2min^{0.5}$, so they belong to the W2 category. Mortars 20PC-80CFA belong to the W1 category.

The flexural strength of the mortars as a function of the amount of CFA residue and the curing time is shown in Figure 9. The flexural strength of the control mortars were 7.9 and 7.8 MPa after 7 and 28 curing days, respectively. This indicates that the flexural strength does not change with the curing time. The incorporation of coal fly ash produces a decrease in flexural strength in the same way that the bulk density and the coefficient of water absorption by capillarity data. The incorporation of 20 and 40 wt % of CFA gives rise to mortars with the lowest reduction of this mechanical property, obtaining values of 6.6 and 6.5 MPa after 7 days of curing, increasing to 8.1 and 7.5 MPa after 28 days of curing, respectively. The incorporation of higher quantities of CFA residue results in decreasing trend of the flexural strength with values lower than 5.5 MPa for the samples with 60 and 80 wt % of CFA after 7 and 28 curing days, respectively. The addition of up to 40 wt % of CFA increases in flexural strength with the curing time. Higher CFA additions produce do not any changes in flexural strength with the curing time. This may be due to the good integration of the coal fly ash in the paste, the adequate degree of reaction of the CFA with the PC and the good particle and matrix adherence, presenting a low proportion of stress concentrators. In addition, according to the bulk density and the coefficient of water absorption by capillarity data, the hydration reactions of the paste have not been completed after 7 curing days. After 28 curing days, the pastes contain a greater quantity of hydrated calcium silicates that result in a more compacted and less porous structure. Pores and other imperfections can act as stress concentrators decreasing the flexural strength of the mortars.

The compressive strength of the control mortar containing only Portland cement and the CFA mortars as function of the content of ash and the curing age are shown in Figure 10. It can be observed that the addition of coal fly ash produces a decrease in the compressive strength, except for mortars containing 20 wt % of CFA (80PC-20CFA) that have compressive strength values higher than the control mortar. The addition of the CFA residues hardly changes the mineralogical composition of the mortar specimens, as the XRD and FTIR data showed. The incorporation of CFA produces a decrease in the compressive strength which could be attributed mainly to the bulk density decrease. The incorporation of CFA causes a slight increase in components that are capable of generating calcium silicates hydrated (CSH), which produces a slight increase in the volume of cementitious material in the mortars that incorporate residues. Regarding the influence of the curing time, the mortar after 7 curing days have higher compressive strength, possibly due to a faster reaction between the coal fly ash and Portland cement in the early stages of curing that contributes to an early strength of the mortars. The UNE-EN 998-1, 2010 standard [20] establishes a minimum compressive strength of 6 MPa for its use in joints. Therefore, all sustainable mortars with the addition of coal fly ash meet this design requirement. In addition, masonry rendering and plastering mortars according to their compressive strength are classified into four categories (Table 9). Mortars with 80 wt % of CFA have a compressive strength of 0.8 MPa and 1.50 MPa, after 7 and 28 days of curing time, respectively, so they belong to the CS I category. All other mortars belong to the category CS IV category.

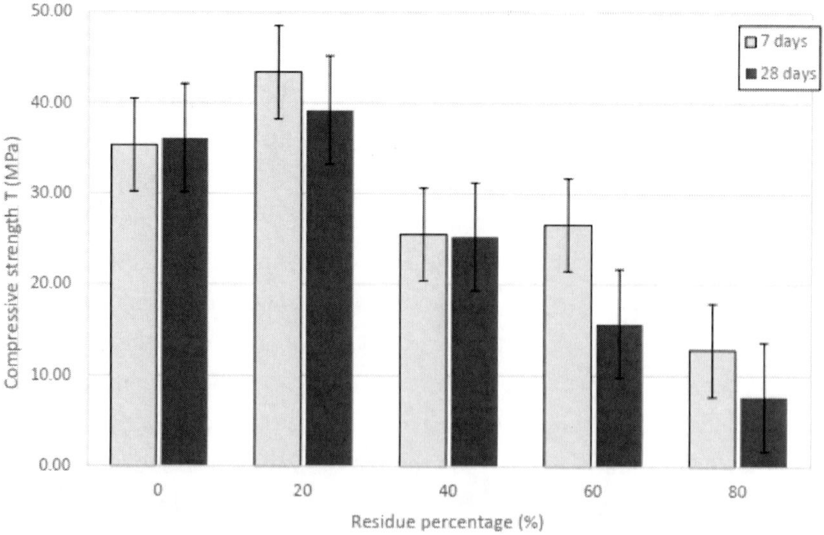

Figure 10. Compressive strength of the xPC-yCFA mortars for 7 and 28 curing days.

Table 9. Masonry rendering and plastering mortars classification according to their compressive strength

Category	Compressive strength range (MPa)
CS I	0.4 – 2.5
CS II	1.5 – 5.0
CS III	3.5 – 7.5
CS IV	≥ 6.0

CONCLUSION

The main conclusions of this study are:

- In the XRD patterns of the pastes the diffraction peaks corresponding to the hydration products of the PC, ettringite and calcium hydroxide, the diffraction peaks corresponding to carbonate and unreacted calcium silicates, main constituents of Portland cement, are observed. This data indicate that after 28 days of curing the hydration process has not been completed yet. The formation of the CHS gel is not observed due to its amorphous nature. As the content of CFA is increased, the intensity of the diffraction peaks corresponding to the calcium hydroxide generated in the hydration reaction of Portland cement decreases due to its reaction with the aluminosilicates present in the ash, producing hydrated calcium silicates. In addition, the diffraction peaks corresponding to the crystalline phases in the residue as quartz are observed.
- The FT-IR spectra of the pastes incorporating CFA and the control mortar are hardly modified although the intensity varies slightly, showing the typical bands of portlandite, gypsum, CHS gel, ettringite and calcite. In the pastes that incorporate ashes the bands assigned to the calcium hydroxide decrease due to its reaction with

the pozzolans. Also a slightly increase of the bands assigned to the CHS gel are observed.

- The study of physical properties indicates that the bulk density of mortars incorporating coal fly ash is hardly modified, due to the similar relative density of the raw materials. The coefficient of water absorption by capillarity increases slightly with the addition of up to 60 wt % of CFA. The coefficient of water absorption by capillary is higher after 7 curing days because the formation of the hydrated calcium silicates have not finished, as indicated by the XRD data.
- The studies of the mechanical properties indicate that the flexural and compressive strength decrease as the content of coal fly ash increases with respect to the control mortar, except for the mortar that incorporates 20 wt% of CFA. The loss of the mechanical properties may be due to the lower reaction of the ash and Portland cement, and to the lower adherence of the ash particles with the matrix of the paste, producing an increase in the capillary porosity of the mortars that incorporate ashes, as indicated by the water absorption by capillarity data, since FRX data indicate that the incorporation of ash causes a slight increase in components capable of generating calcium silicates hydrated (CSH), responsible for the mechanical resistance of mortars.
- 80 PC-20 CFA mortars have mechanical and physical properties similar to the control mortars, 100PC, so its use is justified, from the environmental point of view, by reducing CO_2 emissions, and lead to sustainable mortars.

ACKNOWLEDGMENTS

This work has been funded by the Project "Valuation of various types of ash for the obtaining of new sustainable ceramic materials" (UJA2014/06/13), Own Plan University of Jaen, sponsored by Caja Rural of Jaen. Technical and human support provided by CICT of Universidad de Jaén (UJA, MINECO, Junta de Andalucía, FEDER) is gratefully acknowledged.

REFERENCES

[1] Tosti, L., Zomeren, A. V., Pels, J. R., Comans. R. N. J., Technical and environmental performance of lower carbon footprint cement mortars containing biomass fly ash as a secondary cementitious material. *Resources, Conservation & Recycling* 134 (2018) 25–33.

[2] Directive 2008/98/EC of the European Parliament and of the Council of 19 November 2008 on waste and repealing certain Directives.

[3] CEDEX. *Centro de Estudios y Experimentación de Obras Públicas.* [*Center of Studies and Experimentation of Public Works*] Ministerio para la Transición Económica (2018) [www.cedex.es].

[4] European Coal Combustion Products Association. *Annual Report* 2016. [www.ecoba.org].

[5] Kuroda, M., Watanabe, T., Terashi, N. Increase of bond strength at interfacial transition zone by the use of fly ash, *Cem. Concr. Res.* 30 (2000) 253-258.

[6] Chen, Z., Poon, C. S. Comparative studies on the effects of sewage sludge ash and fly ash on cement hydration and properties of cement mortars. *Construction and Building Materials* 154 (2017) 791–803.

[7] EN 450: *Fly ash for concrete - Part 2: Definitions, specifications and conformity criteria*, 2005+A1:2007.

[8] Haque, M. N., Langan, B. W., Ward, M. A. *High fly ash concrete.* ACI Journal (1984) 54-84.

[9] Tošić, N., Marinković, S., Pecić, N., Ignjatović, I., Dragaš. J., Long-term behaviour of reinforced beams made with natural or recycled aggregate concrete and high-volume fly ash concrete. *Construction and Building Materials*, 176 (2018) 344-358.

[10] McCarthy, M. J., Zheng, L., Dhir, R. K., Tella, G. Dry-processing of long-term wet-stored fly ash for use as an addition in concrete. *Cement and Concrete Composites*, 92 (2018), 205-215.

[11] Tosti, L., Zomeren, A., Pels, J. R., Comans. R. N. J., Technical and environmental performance of lower carbon footprint cement mortars containing biomass fly ash as a secondary cementitious material. *Resources, Conservation and Recycling*, 134 (2018), 25-33.

[12] Liao, K. Y., Chang, P. K., Peng, Y. N., Yang, C. C. A study on characteristics of interfacial transition zone in concrete. *Cement and Concrete Research*, 34 (2004).

[13] Cebrian, J. L., Pisonero, F. Determinación de la superficie específica por el método Blaine en cenizas volantes y cementos puzolánicos. [Determination of the specific surface by the Blaine method in fly ash and pozzolanic cements] *Materiales de construcción* [*Construction materials*] 142 (21) (1971) 81-91.

[14] UNE-EN 1015-11:2007. *Method of test for mortars for masonry. Part 11: determination of flexural and compressive strength of hardened mortar.*

[15] UNE-EN 1015-3:2000. *Methods of test for mortar for masonry: Part 3. Determination of consistence of fresh mortar (by flow table).*

[16] UNE-EN 1015-10:2000 UNE-EN 1015-18:2003. *Methods of test for mortar for masonry. Part 10: Determination of dry bulk density of hardened mortar.*

[17] UNE-EN 1015-18:2003. *Methods of test for mortar for masonry. Part 18: Determination of water absorption coefficient due to capillary action of hardened mortar.*

[18] Agarwal, S., Pozzolanic activity of several siliceous materials. *Cement and Concrete Research,* 36 (2006), 1735-1739.

[19] Rai, U. S., Singh, R. K. Effect of polyacrylamide on the different properties of cement and mortar. *Material Science and Engineering: A*, 392 (1) (2005) 42-50.

[20] EN 998-1:2010. *Specifications of mortars for masonry. Part 1.*

In: Coal Fly Ash
Editor: Miriam Gray
ISBN: 978-1-53614-511-3

Chapter 3

The Effects of Using Coal Ash with Hydrated Lime in Hot Mix Asphalt Wearing Course as a Mineral Filler

Reshma Rughooputh and Abdallah Alhadi Khodabux
Faculty of Engineering, University of Mauritius, Reduit, Mauritius

Abstract

An investigation has been carried out in this study to explore the behaviour of using coal ash with hydrated lime as mineral filler in hot mix asphalt, in an attempt to a sustainable solution for the management of coal ash in Mauritius. In this study, the fly ash and bottom ash were incorporated in various percentages (0%, 1%, 2%, 3% and 4%) each with a fix percentage (1%) of hydrated lime as mineral fillers and as partial replacement of fine aggregate in 40/50 grade bitumen. The asphalt concrete mixes were prepared for a fix asphalt content (5.7%) hence were tested to evaluate their engineering properties which encompass moisture susceptibility, rutting and temperature susceptibility. Even environmental impacts while being in asphalt was evaluated. Mixes modified with the fillers were found to have improved temperature susceptibility and rutting, also showed lower moisture susceptibility and the leachates detected were within acceptable range. It is therefore concluded that both fly ash and bottom ash can be added successfully in asphalt up to 4% each with 1% hydrated lime to produce an asphalt concrete with better serviceability while being safe to the environment.

Keywords: asphalt concrete, coal fly ash (CFA), coal bottom ash (CBA), hydrated lime (HL), moisture susceptibility, mineral filler, and mechanical properties

1. INTRODUCTION

In view of the predictable rise in consumption of coal for power generation in the forthcoming years, ash disposal may become a major environmental problem unless appropriate measures are adopted. The Technical Advisory Committee explained that coal ash can be used as a recyclable by-product under environmental supervision for both encapsulated and unencapsulated purposes [MESD, 2009]. In this chapter, the coal ash has been encapsulated inside the bitumen matrix hence it can help to reduce the need for landfill space and new landfills.

Physically, CFA consisted generally of smoothed surface spherical particles while CBA has a friable popcorn like morphology with porous and angular finish. Both were grey in colour. Moreover HL composed of fine white powder of hexagonal plate crystal structures. Conforming to ASTM C618-85 the CFA was classified as CLASS F as the coal was sourced from South Africa.

Experimental investigations on rutting by Bjorn (2000), Alatas and Ethem (2012) and Gunalaan (2013) have shown that air voids had decreased with increased percentage of HL, CFA and CBA respectively while increasing the Marshall Stiffness of the compacted asphalt. They concluded that the asphalt had become more resistant to permanent deformation. Jiabin (2010) recommended that more than 70% of the fillers should pass the 75μm sieve because the finer the ash the more likely it would be in suspension in the binder forming a mastic contributing to stiffness of the mix rather than settled in the binder. Amelia and Abtahi (2014) investigated the moisture damage of asphalt containing hydrated lime. They reported that the asphalt had become more resistant to moisture damage because it reacted mainly with the carboxylic acid of the bitumen binding the asphaltenes to the aggregates only in presence of water. Hanaa and Hayder (2014) and Scott (1978) observed a decrease in moisture susceptibility of compacted asphalt mixes containing HL and CFA respectively by densification.

Certain materials, such as carbon, are naturally hydrophobic, whereas the oxides and silicates that make up the bulk of fly ash are naturally hydrophilic (Eisele and Kawatra, 2002). They were prone to detachment in presence of water by repelling carboxylic acids and attracted hydrogen molecules from water. Hence causing stripping. Therefore Liao (2007) recommended the use of HL with such type of coal ash. Albayati (2011) and Saad and Ibtihal (2014) reported that the asphalt containing the HL and the CFA respectively may be less susceptible to temperature.

Therefore the objectives of this laboratory investigation were to assess the effect of coal ash with hydrated lime on the resistance of HMA`s wearing courses to moisture and temperature susceptibility, moisture damage and rutting. Over and above that, an environmental evaluation was also performed based on reliable leaching test method.

2. Materials and Methodology

2.1. Materials

The coal ash was obtained from a local power plant and the HL was also obtained locally. The CBA was crushed several times using the Los Angeles Abrasion machine until, at least 75% passed the 75μm sieve so as all the mineral fillers conformed to AASHTO M17. The natural fine aggregate used was 0/3 and 3/6 mm size while coarse was 6/10 and 10/14 mm size. A 40/50 grade bitumen was used.

2.2. Laboratory Test and Standards

2.2.1. Particle Size Distribution

The particle size distribution was carried out for fine, coarse aggregates and for the mineral fillers as well with the test method BS EN 933-1:2012.

2.2.2. Specific Gravity

It was determined for the filler materials according to BS EN 1097:5-7: 2008.

2.2.3. Flakiness Index

The flakiness index of coarse aggregates were determine using BS EN 933-3:2010.

2.2.4. Los Angeles Abrasion Test

The abrasion test was performed on coarse aggregates conforming to BS EN 1097-2:2010.

2.2.5. Preparation of Mix Specimens

The samples were prepared using Marshall Mix Design Method conforming to BS EN 12697:2012. About 500g of bitumen was heated about 160°C to fluid condition in a 2L capacity metal container. The fillers and aggregates were placed in oven at 110 ± 5°C for drying overnight. The correct quantities of dried fillers, aggregates and heated bitumen were placed in a container and heated up to 160°C for 2 hrs. The sample was then poured into a preheated malaxor at 160°C hence was mixed for 4-5 minutes. The mixtures were then poured into 6 separated moulds for compaction using 75 blows of the Marshall compactor on each side of the specimens.

The replacement percentages were 0, 1, 2, 3 and 4% of CBA and CFA each with additional 1% of HL by total weight of the mix. On top of that separated bituminous mastics (bitumen and filler only) using the same stated proportion of fillers were also prepared.

2.2.6. Needle Penetration Test and the Ring and Ball Test

The penetration values of pure bitumen and bituminous mastic were evaluated in accordance to BS EN 1426:2007. The softening points of pure bitumen and bituminous mastic were evaluated in accordance to BS EN 1427:2007. Both tests were used to determine the Penetration Index for both pure bitumen and bituminous mastic which in turn their respective stiffness would be determined using the Van Der Poel nomograph.

2.2.7. Marshall Test

This empirical test has been used to evaluate the different mixture at a fix bitumen content and the parameters of the 3 compacted specimens considered are stability, flow value, bulk density, air voids, voids in mineral aggregates, voids filled with bitumen all conformed to BS EN 1297:2012.

2.2.8. Standard Marshall Immersion Test

This test was performed on the next 3 compacted asphalt specimens for moisture susceptibility determination in accordance to ASTM D1075.

2.2.9. Boiling Water Test

The moisture damage characteristic of loose asphalt mix was determine using this test in accordance to ASTM D3625.

2.2.10. Leaching Test

The leachates extracted from loose asphalt were analysed in accordance to BS EN 12457:4:2002.

3. Results and Discussions

3.1. Particle Distribution Size

It was observed that the fly ash had particle sizes ranging from 0.01-0.5 mm due to the fact that 100% passed sieve size 0.5 mm. The S-curve of particle size distribution was not been able to obtain due to lack of availability of sieve sizes below 0.063 mm. More than 70% passed 0.075 mm sieve showing that fly ash was very fine and was categorised as silt based on the British Standard Soil Classification. The finely crushed CBA was similar to fly ash hence it was also classified as silt. Same was for hydrated lime.

The gradation curves for the coarse and fine aggregates are shown in Figure 2.

3.2. Specific Gravity

Bottom ash possessed the lowest specific gravity of 1.8 due to the fact that it contained porous particles. Hydrated lime has the highest specific gravity of 2.3 because of the crystalline structure which absorbed water and sink at the bottom. For fly ash it has a specific gravity of 2.2 higher than CBA because water was absorbed into the hollow spherical structures it possessed.

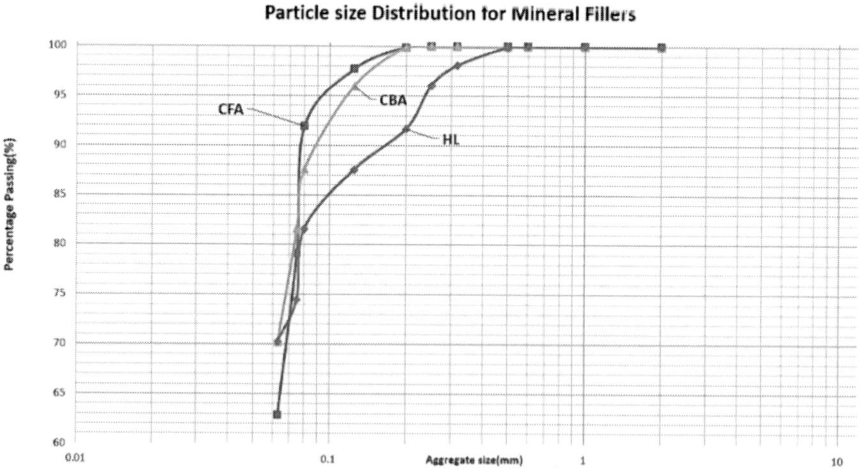

Figure 1. Graphical Representation of particle size distribution of mineral fillers.

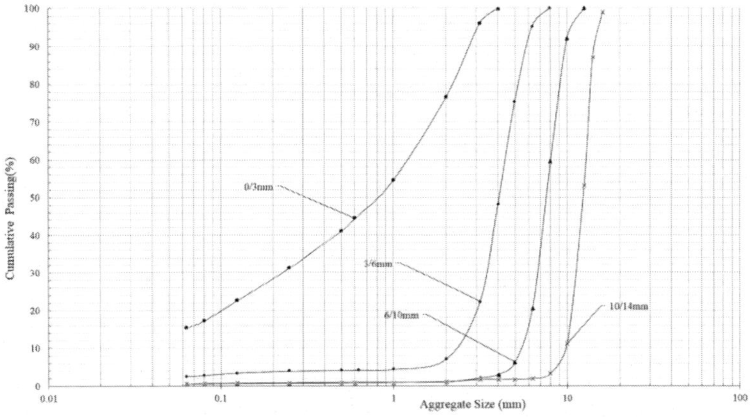

Figure 2. Graphical Representation of particle size distribution of aggregates.

3.3. Flakiness Index and Los Angeles Abrasion Test

The Flakiness Index and Los Angeles abrasion value for coarse aggregate were 8.5% and 30.4% respectively.

3.4. Needle Penetration Test and the Ring and Ball Test Results

3.4.1. Penetration Value

Based on observation, penetration values were decreasing while fillers were added gradually. This significant decrease was due to the gradual increase in viscosity of the bituminous mastic due to progressively increase in the degree of peptisation in the binder medium. Hence it provided resistance to the downwards motion of the penetrating needle into the mastic. Therefore lesser penetration values were obtained same as Albayati (2011).

By distinction, bottom ash as filler produced higher penetration values than fly ash. Bottom ash could be coarser than fly ash, hence bottom ash would likely to settle causing a gradual decrease in viscosity compared with a bottom ash mastic which had been freshly poured into a penetration dish. Upon settling the degree of peptisation decreased causing less resistance to the penetrating needle resulting higher penetration value.

3.4.2. Softening Point

It was observed that the softening point was higher than the control due to addition of fillers. It was increasing significantly with progressively increase in quantity of added fillers. It was attributed to the development of stable dispersion of the fillers as colloidal particles in the binder called peptisation, so greater surface area of the fillers were exposed to adsorption resulting a tightly packed arrangement. Hence it took longer time to soften the mastic compared to pure bitumen. Same results were obtained by Lee (1964).

By contrast the fly ash`s softening points were higher than that of bottom ash. This could be argued that fly ash possessed much finer particles than bottom. The degree of peptisation was greater with fly ash resulting a much stiffer mastic. Hence it took longer time to soften the mastic resulting an increase in temperature.

3.4.3. Penetration Index

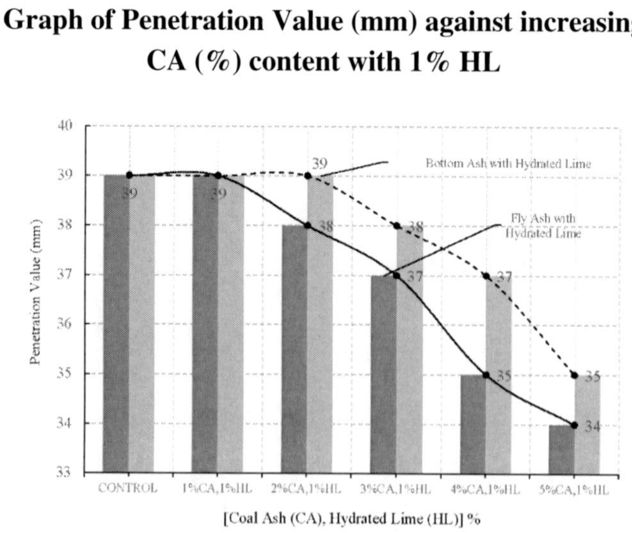

Figure 3. Graph showing trend of penetration value with increasing CA content.

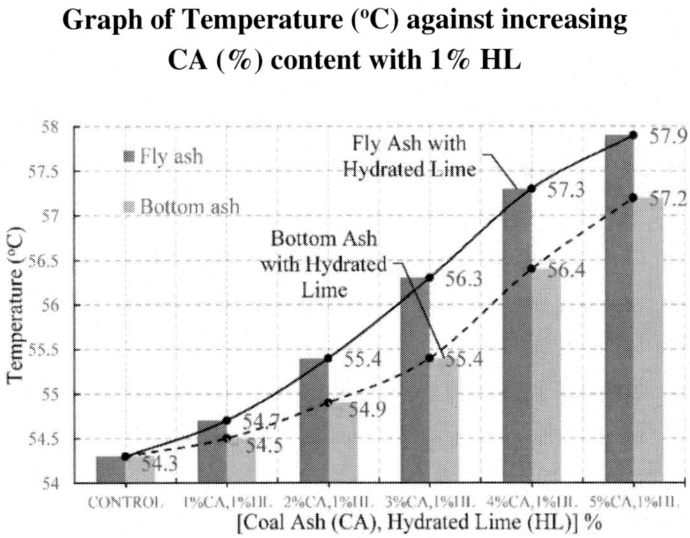

Figure 4. Graph showing trend of softening point with increasing CA content.

The PI increased with addition of mineral fillers. This rise was due to the stiffening effects of the fillers in the binder making it less susceptible to temperature because increased in PI was due to the decrease in penetration value and increase in softening temperature.

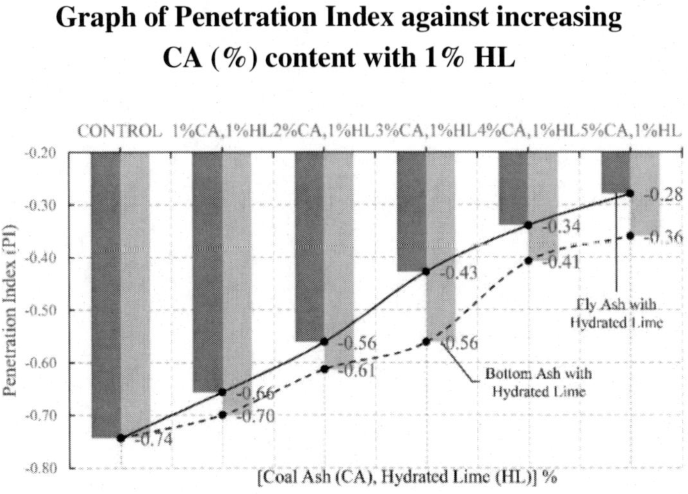

Figure 5. Graph showing trend of PI with increasing CA content.

By contrast, bituminous mastic containing fly ash with hydrated lime resulted higher indexes simply because it was stiffer than bituminous mastic containing bottom ash.

3.4.4. Stiffness Modulus

It was observed that instantly upon addition of mineral filler, its stiffness was increasing, and resulting a hard bituminous mastic which could easily resist rutting hence achieved one of the objectives. The more fillers added the stiffer became the mastic. This significance increase was due to the increasingly degree of peptisation in the mastic. Hence more adsorption of bitumen occurred onto the fillers` surface causing surface energy of attraction between the filler particles which in turn resulted a tightly packed matrix.

By contrast, bituminous mastic containing fly ash resulted higher stiffness modulus compared to mastic containing bottom ash as filler. The reason was due to the finer particles of the fly ash.

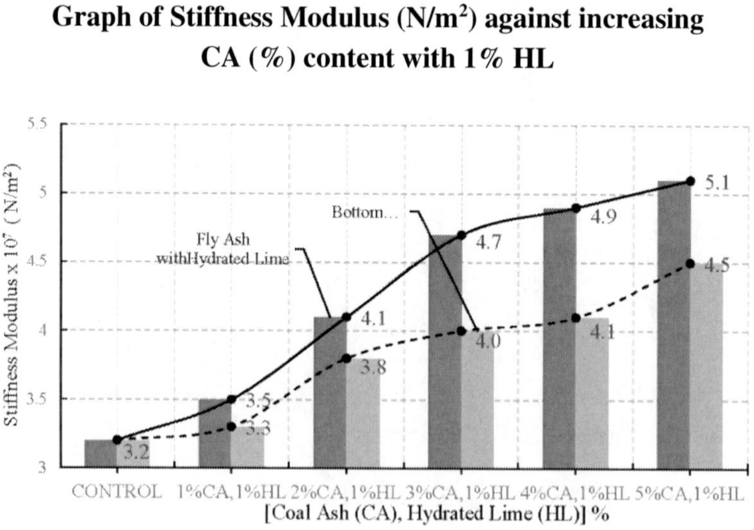

Figure 6. Graph showing trend of stiffness modulus with increasing CA content.

3.5. Marshall Test Results

3.5.1. Bulk Density

The bulk density was increasing upon gradual addition of mineral fillers. This was attributed to the decreased in air voids inside the compacted specimens which had been filled by them thereby making the asphaltic concrete more impervious.

By contrast from the above figure, bituminous mastic containing fly ash with hydrated lime as mineral filler resulted higher bulk density. This was due to the fly ash particles which were more dispersed in the mixture as colloids, contributed to fill more voids between the aggregates compared to bottom ash which were less dispersed due to smaller proportion of fine particles (< 63μm).

Graph of Bulk Density (Mg/m³) against increasing CA (%) content with 1% HL

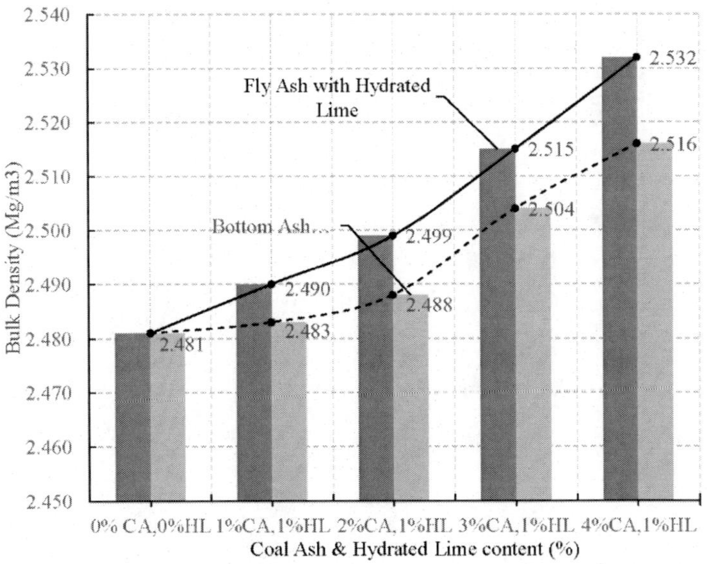

Figure 7. Graph showing trend of bulk density with increasing CA content.

3.5.2. Voids (Va)

It was observed that air voids were decreasing with further increment in percentage of CA. The reason was due to the fact that the fillers extended the effective or free bitumen to occupy a larger volume between the aggregates thereby reducing the volume of air pockets under the same compaction effort.

The results with fly ash were higher than bottom ash because as stated by Tunnicliff (1962), more fly ash particles formed part of the binder as colloids due to higher degree of fineness than bottom ash which greatly formed part of the aggregates due to lesser degree of fineness. Hence fly ash contributed more in diminishing air pockets between aggregates than bottom ash.

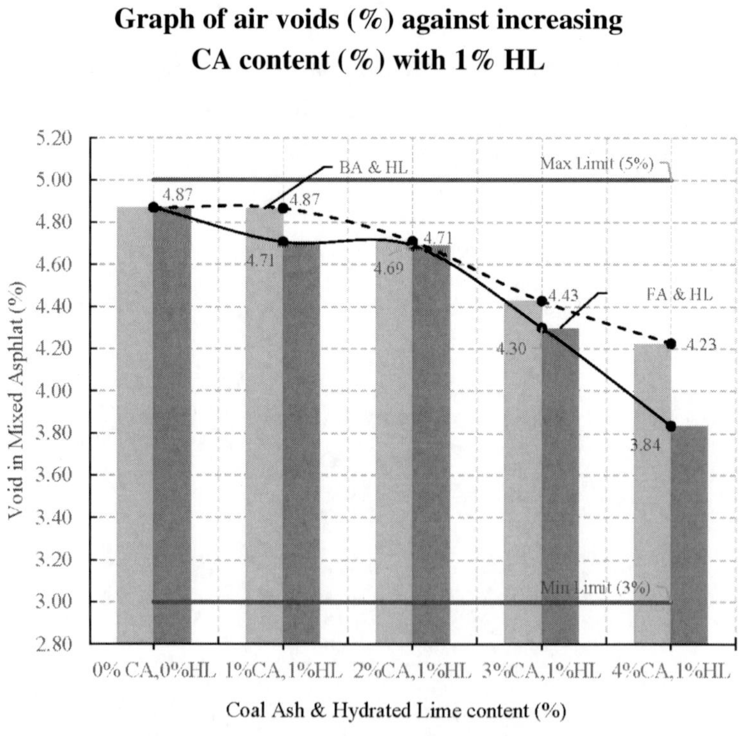

Figure 8. Graph showing trend of air voids with increasing CA content.

3.5.3. Void in Mineral Aggregates

Since VMA consisted of the void occupied by the effective binder and the air voids between the spaces available between the aggregates, hence with a fixed percentage of binder but with a gradual decrement of air voids due to a progressively added mineral fillers would therefore resulted a decrease in VMA.

By contrast, since fly ash resulted a greater reduction in air voids therefore VMA values were smaller compared to bottom ash. Moreover due to the ball bearing effects of the fly ash particles the aggregates were brought closer.

3.5.4. Void Filled with Bitumen

Since VFA was the ability of the bitumen to occupy voidage between the aggregates, hence with the increment of fillers in the mix, caused the binder to extend over larger volume with the filler as colloids. In order to achieve that, the binder squeezed among the filler particles upon compaction thereby extended to cover more space among aggregates, explained Bjorn (2000).

The results were less with bottom ash because lesser particles were in suspension in the binder, hence lessen the extension of the bitumen to occupy more space.

3.6. Standard Marshall Immersion Test Results

3.6.1. Marshall Stability

Instant increase in both unconditioned and conditioned Marshall Stabilities were observed. According to Rebala et al. (1994), the reason was due to the fact that filler particles which were tightly packed together already 'closed-up' the compacted specimens readily forming a highly friction effective bituminous mastic. Hence under load, these filler particles along with coarse and fine aggregates in the bituminous mastic were brought closer to each other causing quick repulsion due to early particles-particles contact resulting a rise in friction which in turn yielding higher internal stabilities. Also the crystalline platelets of the lime provided interlock among the spherical particles of the fly ash.

The conditioned Marshall stability for all samples compared to their respective unconditioned Marshall stability was decreasing when submerged in water at 60°C for 24 hrs. It was due to the fact that diffusion of water through the specimens under the influence of temperature, caused swelling due to internal voids expansion thus diminishing the cohesive strength of the specimens. Moreover ingress water at the interface of the aggregates and the binder caused a decrease in the adhesive strength.

By contrast, fly ash resulted higher stability than bottom ash because of greater stiffness of the mastic and had more internal friction contributing to the internal stability the compacted mix. In case of the decrease in the conditioned Marshall stability of all samples, it will be discussed in the retained stability sub-section.

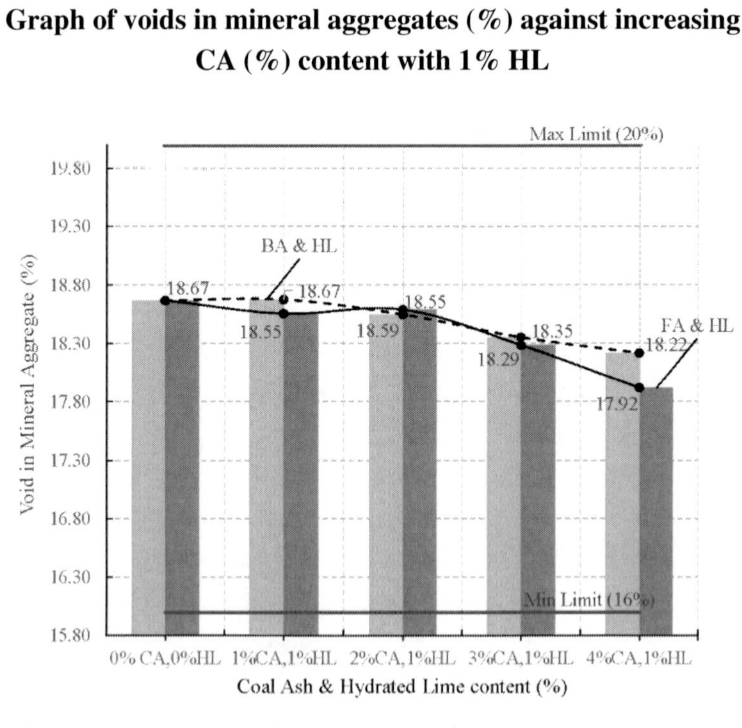

Figure 9. Graph showing trend of VMA with increasing CA content.

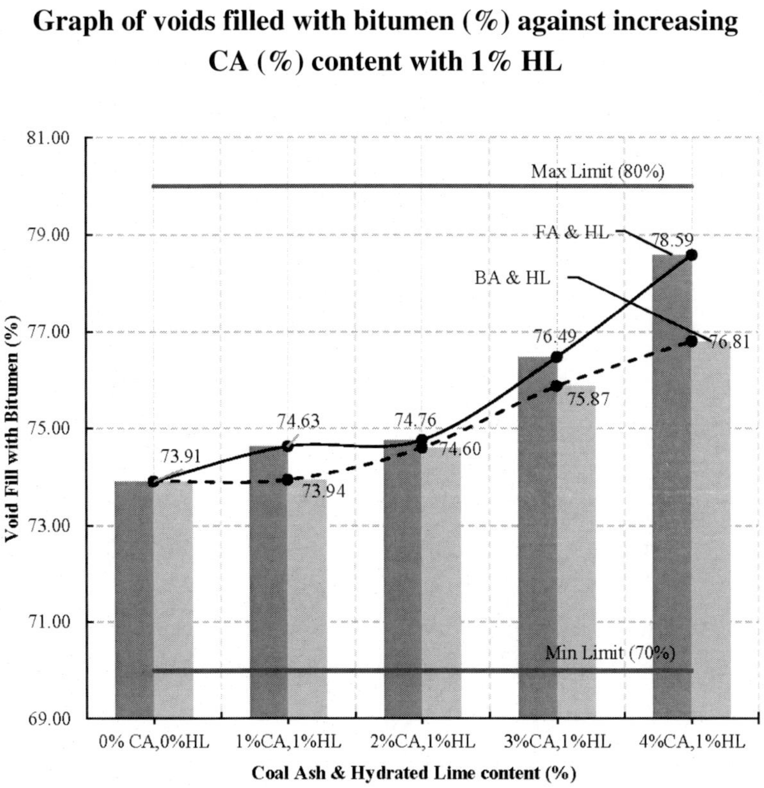

Figure 10. Graph showing trend of VFA with increasing CA content.

3.6.2. Flow

Based on observation, unconditioned flow was decreasing with gradual addition of the mineral fillers. This was due to the rise in stiffening of the mixtures. Moreover the bituminous mastic coated the aggregates further inhibit deformation of the compacted specimens by providing greater internal friction between the aggregates.

The conditioned flow values were much higher than the unconditioned samples because the mastic had been weakened or softened on purpose of the test due to diffusion of hot water through the submerged mastic for 24 hrs. hence the effect of friction inhibiting deformation was decreased but it should not decrease to such a degree making the sample susceptible to moisture.

However the conditioned flow increased at a certain point for both fly ash and bottom ash bituminous mastic. It was attributed to the weakening of the bituminous mastic under action of temperature. But due to the rise in stiffness due to further addition of filler compensated that increased in flow. Hence, resisting the weakening of the binder due to temperature.

3.6.3. Retained Stability

The conditioned Marshall Stability was decreasing for all samples even for control because of diffusion of water through the compacted specimens leading to diminish the cohesive strength of the binder and for those which had successfully attained to reach the interface of aggregates and the binder caused weakening of adhesive strength of the binder to the aggregates. However such decrease in strength was then used to assess how much the specimens could retain the strength after being affected by water.

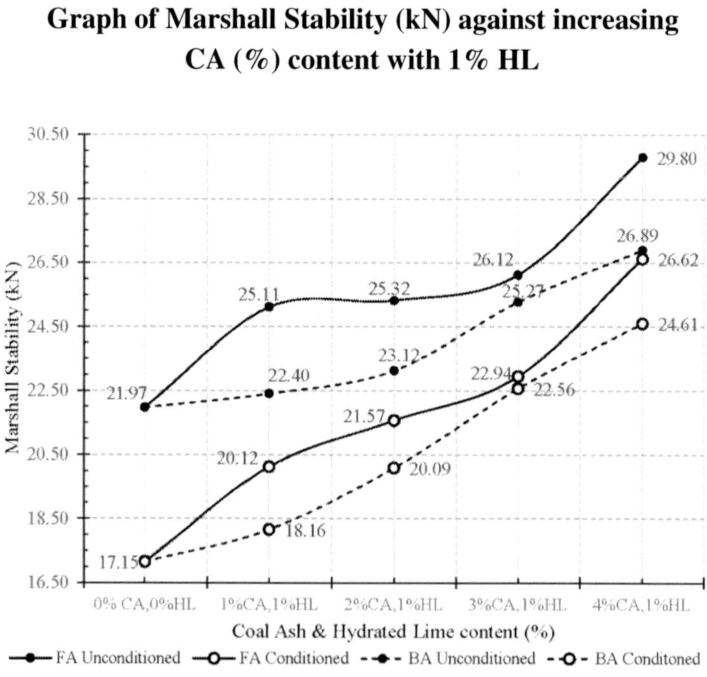

Figure 11. Graph showing trend of Marshall Stability with increasing CA content.

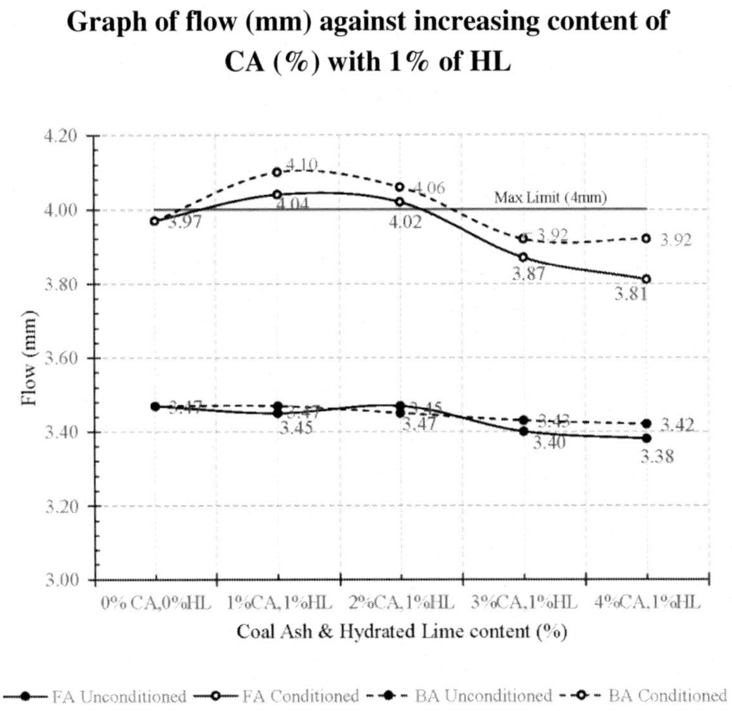

Figure 12. Graph showing trend of flow with increasing CA content.

It was observed that the Retained Marshall stability was increasing with progressively added mineral fillers. This was due to the fact that the amount of air voids were very few with the added fillers hence reducing the probability of the voids in a compacted specimen to be interconnected which could ease the ingress of water inside the specimen. Since ingress of water was reduced, the diffusion of the water through the compacted samples was also reduced, diminishing the probability of the diffused water to permeate at the interface of the aggregate and the binder which would contribute to weaken the adhesive strength of the binder.

Moreover the damage to the cohesive strength of the mastic due to the ingress of water was also reduced by the gradual addition of filler into the mix. By contrast it was observed that bottom ash with hydrated lime resulted higher retained stability. It was due to the high carbon content the bottom ash possessed hence making it more attractive to bitumen rather than water.

Since both the fly ash and the bottom ash were hydrophilic, they were prone to moisture damage hence hydrated lime would react with the permeated water through voids and channels to form an insoluble calcium ions coating to the ash particles increasing their adhesive strength to the binder. However the effect was decreasing with increasing percentage of coal ash content compared to the lime content.

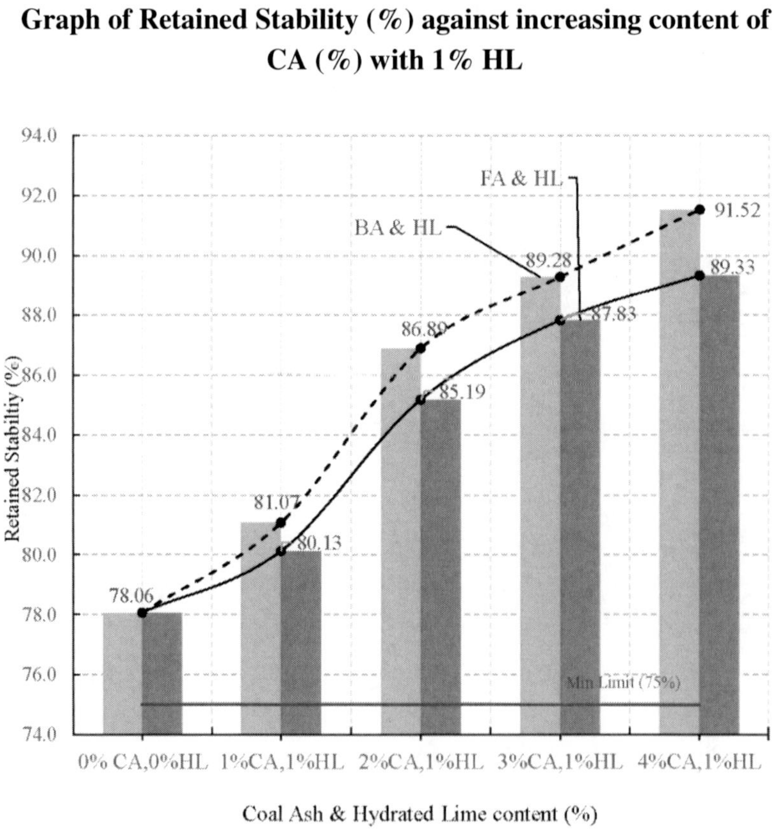

Figure 13. Graph showing trend of retained stability with increasing CA content.

3.7. Boiling Water Test Results

Plate 1. Boiling Test result for control mix.

Observation

- Almost all aggregates were coated except a few on which their original colour was exposed due to loss of adhesion of the binder.
- Some fine particles came loose but most were held on surface of coarse aggregates proving presence of cohesive bond.
- Stripping: Low

Plate 2. Boiling Test result for 1% FA, 1% HL.

Observation

- All aggregates were coated proving adhesion of the binder.

- Few fine particles came loose but some were held on surface of coarse aggregates proving existence of cohesive bond (oily aggregates).
- Stripping: Low

Plate 3. Boiling Test result for 2% FA, 1% HL.

Observation

- Observation was same as previous one
- Stripping: Low

Plate 4. Boiling Test result for 3% FA, 1% HL.

Observation

- All aggregates were still fully coated no loss in adhesion was observed.

- The asphaltenes and maltenes (oily substances) decreased resulted the fine particles to come loose. Only a few remained attached on coarse aggregates due to the binder.
- Stripping: Low

Plate 5. Boiling Test result for 4% FA, 1% HL.

Observation

- All aggregates were still fully coated no loss in adhesion was observed.
- The asphaltenes and maltenes (oily substances) were almost remove resulting almost all fine particles to come loose leaving behind the grey colour of the ash adhered to the aggregate due to the binder.
- Stripping: Slight

Plate 6. Boiling Test result for 1% BA, 1% HL.

Observation

- All aggregates were perfectly coated no loss in adhesion was observed.
- The asphaltenes and maltenes were still available to hold some of the fine aggregate on coarse aggregates` surface but some came loose.
- Stripping: Low

Plate 7. Boiling Test result for 2% BA, 1% HL.

Observation

- Observation was same as previous one
- Stripping: Low

Plate 8. Boiling Test result for 3% BA, 1% HL.

Observation

- The bitumen had stripped from the bottom ash leaving behind the grey colour of the ash but adhesion of the binder to aggregate remained.
- Due to the stripped binder from the ash loss of cohesion in the binder was noticed due to the loose fine aggregates.
- Stripping: Slight

Plate 9. Boiling Test result for 4% BA, 1% HL.

Observation

- All aggregates were fully coated so no loss of binder from aggregates` surface.
- Some fine particles came loose.
- Stripping: Low

Since the coal ash was hydrophilic, a progressively added amount resulted the binder to strip off them but not onto the aggregates hence proved they were really hydrophilic. Since hydrated lime is still the world best anti-stripping filler, it could be seen that with high ratio of lime to coal ash (1:1) contributed to prevent the stripping of the binder from the coal ash.

Table 1. Concentration of trace elements from eluate analysis

Trace Elements	Highest Concentration (mg/kg)			
	Sample 1 (4% FA with 1% HL)	Sample 2 (4% BA with 1% HL)	Mauritius Effluent Discharge Standards	Mauritius Drinking Water Standards
Aluminium	0.74	0.53	5	0.2
Arsenic	N/D	N/D	0.1	0.01
Cadmium	N/D	N/D	0.01	0.003
Cobalt	N/D	N/D	0.05	No std
Chromium	N/D	N/D	0.05	0.05
Copper	N/D	N/D	0.5	1
Manganese	N/D	N/D	0.2	No std
Nickel	N/D	N/D	0.1	0.02
Lead	N/D	N/D	0.05	0.01
Zinc	0.24	0.38	2	3

(N/D: Not Detected)

3.8. Leachability Test Results

Only traces of Aluminium and Zinc were detected due to the fact that the coal ash were highly concentrated of these two metals. Even as encapsulated trace elements were still detected because the samples were harshly agitated for 24 hrs hence overestimating a real road scenario. Zinc for both samples were far below the Mauritius Effluent Discharge Standards (MEDS) and Mauritius Drinking Water Standards (MDWS). Aluminium for both samples respected the MEDS but not the MDWS.

CONCLUSION

The results obtained from the laboratory investigation confirm the general findings from previous research that adding CBA and CFA with HL to asphalt increase Marshall Stability value and decrease the rutting and temperature susceptibility potential. In addition, as expected, a decrease in voids upon addition of the fillers to the asphalt was registered due to densification of the compacted mix.

The moisture susceptibility potential also decreased due to an increase in peptisation of the filler portion but for a short period of time. It is therefore recommended perform a Modified Marshall Immersion Test to directly test its durability for prolonged period of time in water heated at 60°C.

REFERENCES

Alatas, T. and Ethem, K., 2012. The effects of using Styrene-Butadiene-Styrene and Fly Ash Together on the Resistance to Moisture-Induced Damage, Permanent Deformation and Fatigue of Hot Mixture Asphalt. *Journal of Civil Engineering*, Vol. 17(5), pp.1030-1035.

Albayati, A.H., 2011. Mechanistic Evaluation of Lime-Modified Asphalt Concrete Mixtures. *Journal of Engineering*, Vol.17, pp.1447-1458.

Bjorn, O., 2000. *Ageing and Degradation of Asphalt Pavement on Low Volume Roads*. Thesis (PhD). The Norwegian University of Science and Technology.

British Standard Institution, 2004. Test methods for hot mix asphalt. *BS EN 12697*, British Standard Institution, London.

Eisele, T. C. and Kawatra, S. K. 2002. *Use of froth flotation to remove unburned carbon from fly ash*, 23, pp. 1- 10. DOI: 10.1080/08827500290110188.

Gunalaan, V., 2013. Performance on Coal Bottom Ash in Hot Mix Asphalt. *International Journal of Research in Engineering and Technology*, Vol. 2, pp.25-32.

Hanaa, M. and Hayder, A., 2014. Laboratory Examination for the Effects of adding Hydrated Lime on the Moisture Damage Resistance of Asphalt Concrete Mixtures. *Journal of Engineering*, Vol. 5, pp. 11-12.

Jiabin, F., 2010. *Challenges to Increase Use of Coal Combustion Products in China*. Thesis (MEng). Linkopings Universitet.

Liao, M.C., 2007. *Small and Large Strain Rheological and Fatigue Characterisation of Bitumen – Filler Mastics*. Thesis (PhD). University of Nottingham.

Ministry of Environment & Sustainable Development (MESD), "*Technical advisory committee on coal ash management*," Final Report, 2009.

Rebala S., Estakhri C. K., Gownder M. and Little D., 1994. *Short-term Guidelines to improve crumb rubber modified asphalt concrete pavements.* Texas Transportation Institute. The Texas A&M University System College Station, Texas 77843-3135. Research Report 1332-2F.

Scott J. A., 1978. Adhesion and Disbonding Mechanisms of Asphalt used in Highway Construction and Maintenance. *Journal of the Association of Asphalt Paving Technologists*, 47.

Saad I. S. and Ibtihal M. L., 2014. Impact of Asphalt Additives on Rutting Resistance of Asphalt Concrete. *International Journal of Scientific Research in Knowledge, (IJSRK)* Vol. 2(3), pp. 151-159, Malaysia.

Tunnicliff D. G., 1962. A Review of Mineral Filler. *Proceedings of the Association of Asphalt Paving Technologists*, 31: 118-150.

In: Coal Fly Ash
Editor: Miriam Gray
ISBN: 978-1-53614-511-3

Chapter 4

SYNTHESIS OF NANOCRYSTALLINE ZEOLITE NA-A FROM COAL FLY ASH AND DIFFERENT ALUMINUM SOURCES: APPLICATION FOR HEAVY METALS REMOVAL

Denise A. Fungaro[1,*], Juliana C. Izidoro[1], Sabine. N. Guilhen[1], Katia C. Silva[1] and Shaobin Wang[2]
[1]Instituto de Pesquisas Energéticas e Nucleares, IPEN–CNEN/SP, São Paulo, SP, Brazil
[2]Curtin University, Department of Chemical Engineering, Perth, WA, Australia

ABSTRACT

This chapter describes the synthesis process to obtain zeolite Na-A from coal fly ash via fusion with NaOH followed by hydrothermal reaction. Coal fly ash is a low cost source of both silica and alumina. The molar ratio of SiO_2/Al_2O_3 for zeolite 4A formation was achieved by the addition of sodium aluminate solution or aluminum waste from tertiary industry as aluminum sources. The effects of temperature and time at the fusion step and the hydrothermal reaction time on crystalline products were investigated. The synthesized products were studied for their purity and yield as well as were characterized by different techniques such as X-ray powder diffraction (XRD), scanning electron microscopy (SEM) and cation exchange capacity (CEC), among others. Analysis of the properties of the raw materials and the products demonstrates that the by-products can be used to produce a high purity zeolite A. The optimum conditions related to the synthesis was set and the samples were pure, highly crystalline and white. The synthesized zeolites were used for both the removal of Cd (II) and Zn (II) from aqueous solution. The results showed that coal fly ash and aluminum source can be transformed into zeolite A, which is considered as value-added material with adsorption properties.

1. INTRODUCTION

1.1. Zeolite Synthesis from Fly Ash

Coal fly ash generated during the coal combustion process in the thermal power plants can be used as a raw material in the synthesis of other materials with a high added value as a way to alleviate the environmental problems related to its disposal.

Some characteristics of the coal ashes, such as the presence of silicon and aluminum as major constituents in their chemical composition (Fungaro et al., 2013), allow these materials to be used in the synthesis of adsorbent materials, such as zeolites.

Synthesis of high purity zeolites using coal ash has been reported in academic papers in recent years because this kind of product has several applications in the industry, besides it can be synthesized from an abundant and cheap residue.

Mallapur and Oubagaranadin (2017) presented a complete review of zeolite synthesis by using different types of residues, and the most commonly used residue for this purpose is the coal ash.

Several types of zeolites can be synthesized from coal ash, such as zeolites X, Y, A, P, ZSM-5, ZK-5 and W (Mallapur & Oubagaranadin, 2017), most of which have been used in commercial applications (Petrov & Michalev, 2012).

Among the different types of zeolites, Na-A zeolite is one of the most important zeolites in terms of research interest and for studies on the commercial scale because it's high cation exchange capacity and varied applications (Querol et al., 2002; Ríos et al., 2009).

Na-A Zeolite ($Na_{12}(AlO_2)12(SiO_2)_{12}.x\ H_2O$) is characterized by its three dimensional cubic structure and the pore opening of 0.42 nm. The silica to alumina ratio in zeolite A equals to 1 and the numbers of mobile Na cations and Al atoms are equal. Thus, it presents relatively higher exchange capacities than others zeolites.

The ashes used for zeolite A synthesis can be the fly ashes (Chang & Shih, 2000; El-Naggar et al., 2008; Panitchakarn et al., 2014; Volli & Purkait, 2015) or the bottom coal ashes (Fansuri et al., 2010; Chareonpanich et al., 2011). Fly ashes have been more used because they are produced in greater quantity in the thermal power plants.

Synthesis of zeolite A using the coal combustion by-products can be done with the pretreatment of ash to remove impurities (Panitchakarn et al., 2014), with external aluminum source addition (Petrov & Michalev, 2012), with fusion process (Wang et al., 2008; Volli & Purkait, 2015; Hong et al., 2017), or with the Si and Al solution extracted from the ashes (Ye et al., 2008).

Ojumu et al. (2016) investigated the possibility of replacing high temperature fusion with less energy intensive sonochemical treatment for the synthesis of zeolite A from coal ash. The results showed that fusion step can be replaced by high intensity sonication, making the synthesis of pure phase zeolite A at lower temperatures and reduced times when compared to other processes.

Zeolite A synthesized from coal fly ash is usually characterized by X-ray diffraction, scanning electron microscopy and X-ray Fluorescence techniques (Chang & Shih, 2000; Fansuri et al., 2010; Chareonpanich et al., 2011) and can be used in various applications, such as the removal of mercury and lead from wastewater (Somerset et al., 2007), removal of cobalt ions (Chang & Shih, 2000), removal of cadmium and zinc from aqueous solutions (Izidoro et al., 2013), removal of cesium ions

(El-Naggar et al., 2008), as catalyst for the biodiesel production (Volli & Purkait, 2015), among others.

1.2. Industrial Aluminum Hazardous Waste

Currently, aluminum is a product with a growing consumption due to its remarkable properties which are suitable in most areas of everyday life as transport, engineering, construction, packaging, etc. The worldwide implementation of aluminum has been successful due to the cost reduction in the manufacture process and the large possibilities of recover.

In Brazil, the aluminum production represents an important national commodity since the country is the third largest producer of alumina also the fourth largest producer of bauxite and has the third largest reserves of this ore in the world.

In addition to primary production, recycling rates are also highlighted in Brazil. Since the early 1990's, with the introduction of aluminum beverage cans in the national market, aluminum recycling activities has continuously grown and Brazil is one of the world's largest aluminum recyclers, forward of Japan and Europe. The recycling rate for aluminum beverage cans reached 97.9% in 2015, thereby sustaining Brazil's leading standing in global ranking for over a decade (ABAL, 2017).

The aluminum recycling in Brazil consists of secondary and tertiary aluminum industry (Shinzato & Hypolito, 2005).

The primary industry, which obtains aluminum from treatment of bauxite, produces dross that usually, contains about 80 wt% of metallic aluminum. Due to high content of metal, such slags are reused along with aluminum scrap (beverage cans, flakes of industries that use aluminum as raw material and others) by the secondary companies.

The secondary industry recovers metal by melting it in rotary kilns fueled by combustible oils, using sodium chloride as a flux. This process generates new slag, less rich in aluminum (amount about 20% of total volume), but with high sodium chloride contents, metal oxides, carbides and nitrides.

The management and valorization of aluminum saline slags produced by the secondary aluminum industry have been reported in literature (Tsakiridis, 2012; López-Delgado & Tayibi, 2012; Gil & Korili, 2016).

Finally, the tertiary industry uses different processes to recover the residual aluminum present in the metal aluminum waste of the secondary industry. The recovered aluminum returns to the secondary companies, which refund it with the other scraps and sell the recycled metal to transformation industries.

The wastes from tertiary aluminum, in general, are a rather complex mixture containing free metals, metal oxides, and salts. The disposal of this waste without any treatment can contaminate the surface levels of soils and affect the quality of groundwater and surface waters. The effluent produced has high level of salts, ammonia and is very alkaline, and therefore modifies the pH of these waters and is very toxic to aquatic organisms (Shinzato & Hypolito, 2016).

The suggestion is that the waste from tertiary aluminum industry should be stored in a controlled landfill or reused in some way by remedying the environmental problems caused by this material, especially the contamination of groundwater. An alternative of reuse is employed this waste as aluminosilicate source to synthesize high value-added materials such as zeolite.

In previous studies, Al-containing wastes was used as one of the aluminosilicate precursors for the synthesis of Na-zeolites, namely NaP1, SOD and ANA. For the synthesis of zeolites, the low SiO_2/Al_2O_3 molar ratio from the waste was increased by adding sodium silicate solution (López-Delgado et al., 2014; Sánchez-Hernández et al., 2016; 2017).

1.3. Low Cost Adsorbent

Activated carbon has been widely used as adsorbent in wastewater treatment throughout the world. However, its widespread use is sometimes restricted due to the high costs, since higher the quality of activated carbon, the greater its cost.

The use of low cost absorbent, as zeolite synthesized from industrial wastes, replacing the activated carbons for the treatment of wastewater containing pollutants can to overcome the economic disadvantages of adsorption process.

The applications of fly ash zeolites in water treatment were reviewed by Koshy and Singh (2016).

This chapter presents the synthesis process to obtain NaA zeolite employing coal fly ash and aluminum waste from tertiary aluminum as aluminosilicate source. Experimental synthesis parameters were investigated to determine the optimal conditions to obtain high quality zeolite. Furthermore, the removal efficiency of Zn^{2+} and Cd^{2+} from aqueous solution by zeolitic materials was also discussed.

2. Experimental Section

2.1. Materials

Coal fly ash samples were collected from electrostatic precipitators of Jorge Lacerda Power Plant (JLPP) located in Santa Catarina State and Charqueadas Power Plant (CHPP) located in Rio Grande do Sul State. The ashes were used without any pretreatment and were characterized (Izidoro et al., 2012). Sodium hydroxide (97%) and sodium aluminate (100%) obtained from Sigma-Aldrich Pty. Ltd. (AUS) were used in zeolite synthesis. The residue of aluminum tertiary industry (RA) was provided by the company Latasa Recycling located in São Paulo State, Brazil.

2.2. Zeolite Synthesis from Fly Ash and Commercial Sodium Aluminate

Synthesis of NaA zeolite was executed by a two-step process. Firstly, fly ash (JLPP or CHPP) was mixed with of NaOH at 1:1.2 (wt %) ratio.

The mixture was heated at 500ºC for 1 h. After the cooling at room temperature, commercial sodium aluminate was added to the grinded mixture to control the molar ratio (SiO_2/Al_2O_3 = 1) and mixed with water. The mixture was then stirred at room temperature for 16 h and heated to at 100°C for 7 h. After cooled at room temperature, the suspension was filtered and the solid was repeatedly washed with 1 L of deionized water and dried at 105°C for 16 h. The products obtained were labeled as: ZAJL and ZACH, based on the fly ash sources (Izidoro et al., 2013).

2.3. Zeolite Synthesis from Fly Ash and Residue of Aluminum Tertiary Industry

Zeolite synthesis process by fusion treatment was performed by placing 20 g of CHPP with 40 g of RA and 72 g of NaOH. The ratio (CHPP + RA):NaOH was fixed at 1:1.2 (wt%). The RA was added to adjust the molar ratio of SiO_2/Al_2O_3 to 1 for zeolite 4A formation. The mixture was heated at the selected temperature (500ºC or 600ºC) for different periods of time (1 h and 2 h). The fused sample was allowed to cool to ambient temperature and ground into a fine powder using a mortar and pestle. Then, 600 mL of deionized water were added and the mixture was stirred at room temperature for 16 h (aging step). The resulting slurry formed was subjected to hydrothermal crystallization at 100°C for 7 h. After cooling down to room temperature, the suspension was centrifuged at 2000 rpm for 10 min, the solid was washed with 1 L of deionized water and dried overnight at 80°C. The zeolites formed (ZRA-1 to ZRA-4) was powdered for further use (Izidoro et al., 2013).

2.4. Characterization of Materials

The mineralogical composition of the samples was determined by X-ray diffraction (Bruker-AXS D8 Advance Powder Diffractometer) using Cu Kα radiation at 40 kV and 40 mA. The scan rate was 0.02 /s and ranged from 5 - 80° (2θ). Phase identification was made by searching the ICDD powder diffraction file database, with the help of Diffrac Plus EVA – Bruker files for inorganic compounds. To verify the morphology, samples were covered with a thin layer of gold to make them conductive and examined using a scanning electron microscope (Zeiss - EVO 40 XVP).

The specific surface areas of zeolite samples were determined by a BET Surface Area Analyzer (Micromeritics – Gemini 2360). Prior to determination of the specific surface area, samples were heated at 423.15 K for 12 h to remove volatiles and moisture in a degasser (Micromeritics – VacPrep 061). The BET surface areas were obtained by applying the BET equation to the nitrogen adsorption data.

In cation exchange capacity (CEC) measurements, samples were saturated with sodium acetate solutions ($1x10^{-3}$ mol L^{-1}), washed with distilled water ($1x10^{-3}$ L^{-1}) and then mixed with ammonium acetate solution ($1x10^{-3}$ mol L^{-1}). The sodium ion concentration of the resulting solution was determined by atomic absorption spectrophotometry (Varian – Spectraa-50).

The pH and the conductivity were measured as follows: samples (0.25 g) were placed in 25 mL of deionized water and the mixture was stirred for 24 h in a shaker at 120 rpm (Ética — Mod 430). After filtration, the pH of the solutions was measured with a pH meter (MSTecnopon — Mod MPA 210) and the conductivity was measured with a conductivimeter (BEL Engineering - Mod W12D). The particle size of the materials was measured using a laser based particle size analyser, namely a Malvern MSS Mastersizer 2000 Ver. 5.54.

Leaching and solubilization were tested according to the Brazilian Standard NBR 10005 and 10006, respectively (ABNT NBR 20005-2006, 2004).

2.5. Adsorption Studies Using ZAJL and ZACH

Batch processes for metal ion adsorption were performed in single and binary systems by shaking (Orbital Shaker Incubator – Mod SF-670) at

120 rpm for 24 h. In the experiment, 0.5 g of each zeolite was mixed with 50 mL of metal ion solutions with different and known initial concentrations. The applied solid/liquid ratio was 10 g L^{-1}. The adsorption capacity of the adsorbent per mass unit (q_e) was calculated by Eq. (1). The metal ion adsorption percentage (Ad %) was obtained by Eq. (2) and the total metal ion adsorption percentage (Ad_{Tot}%) was calculated by Eq. (3)

$$q_e = \frac{(C_0 - C_f) x V}{M} \tag{1}$$

$$Ad_{\%} = \left(\frac{C_0 - C_f}{C_o}\right) x100 \tag{2}$$

$$Ad_{Tot\%} = \frac{\Sigma(C_0 - C_f)}{\Sigma C_0} x100 \tag{3}$$

where C_o and C_f are the initial and final metal ion concentrations (mg L^{-1}), respectively, V (L) is aqueous phase volume and M (g) is the weight of adsorbent used.

The data of isotherms were obtained after an equilibrium time of 24 h at 25°C using batch method. After the equilibrium time, the supernatants were separated from the adsorbent by filtration and the concentrations were determined by atomic absorption spectrophotometry (Varian – Spectraa-50). For single metal-ion systems, the initial metal concentration ranged from 653.8 to 1961.4 mg L^{-1} and from 1124.1 to 3372.3 mg L^{-1} for Zn^{2+} and Cd^{2+} ions, respectively. In the binary metal-ion systems, initial concentration of zinc or cadmium was maintained at the lowest concentration to check the influence of competitive ions during the adsorption process under different concentrations.

The linear form of the Langmuir model (Type 1) and Freundlich model were used to determine isotherm parameters and constants and are described in the literature (Tran et al., 2017).

2.6. Adsorption Studies Using ZRA-1

The adsorption was performed using the batch procedure. Equilibrium adsorption experiment was performed using various concentrations of cadmium (438 to 1255 mg L^{-1}) and zinc (152 to 463 mg L^{-1}) solution. A mixture of 0.25 g of ZRA-1 with 25 mL of metal ion solution with various initial concentrations was shaken at 120 rpm for 24 h at 25°C. The collected samples were then filtrate and the concentration in the filtrate solution was analyzed using inductively coupled plasma optical emission spectrometry (Spectro ARCOS, Spectro Analytical Instruments, Kleve, Germany). All experiments were performed in duplicate. The adsorption capacity of the adsorbent and the removal percentage were calculated using the Eq. (1) and (2), respectively.

The data of the isotherm adsorption equilibrium were fitted using nonlinear equations of Langmuir, Freundlich, Temkin and Dubinin-Radushkevich (D-R) models. The determination coefficient (R^2) was used to determine the best-fitting isotherm to the experimental data. In addition to R^2 value, six statistical error deviation functions were examined to evaluate the applicability of each model isotherm equation to the experimental data, abbreviated as ARED (Average relative error deviation), SSE (Sum of Squared Errors), MPSED (Marquardt's percent standard deviation), HYBRID (Hybrid fractional error function), SAE (Sum of absolute error) and X^2 (Chi-square). The isotherm and error functions equations are described in the literature (Ncibi, 2008; Alcântara, 2016).

3. Zeolite A from Fly Ash and Commercial Sodium Aluminate

This part of chapter present an overview of the strategy used for the synthesis and application of fly ash based-zeolite A described in the literature (Izidoro et al., 2013).

3.1. Synthesis and Characterization

3.1.1. Mineralogical Composition

Figure 1 shows the X-ray patterns of pure zeolite A obtained from two different types of Brazilian fly ashes. According to the diffractograms, the products of the synthesis presented only one single phase of pure zeolite. Quartz and mullite peaks, normally found in the diffractograms of fly ash, disappeared due to the fusion step, which can easily dissolve those components for zeolites formation in the later stages of the synthesis.

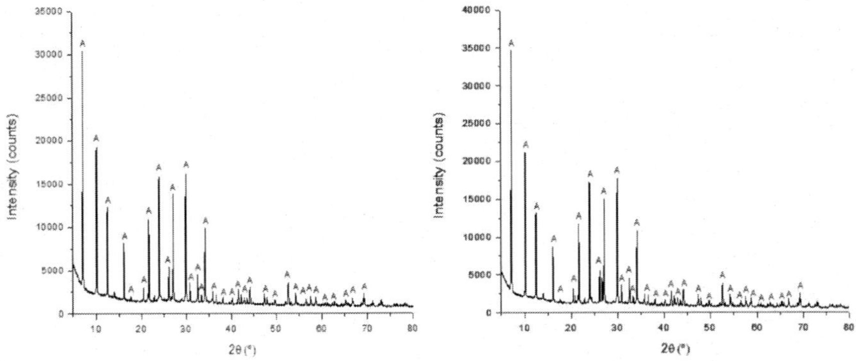

a b

Figure 1. XRD patterns of zeolite A synthesized by two-step process a) ZAJL and b) ZACH (A = zeolite A).

3.1.2. Morphological Analysis

According to literature, zeolite A presents cubic structure (Shiguemoto et al., 1993; Wang et al., 2008; Yaping et al., 2008; Tanaka and Fujii, 2009). Zeolites A synthesized from two different types of Brazilian fly ashes can be clearly seen in Figure 2. According to Figure 2, zeolites A presented different crystals size. The same kind of zeolites can present different crystals sizes due to the different crystal growth rate during the hydrothermal treatment (Hui et al., 2006).

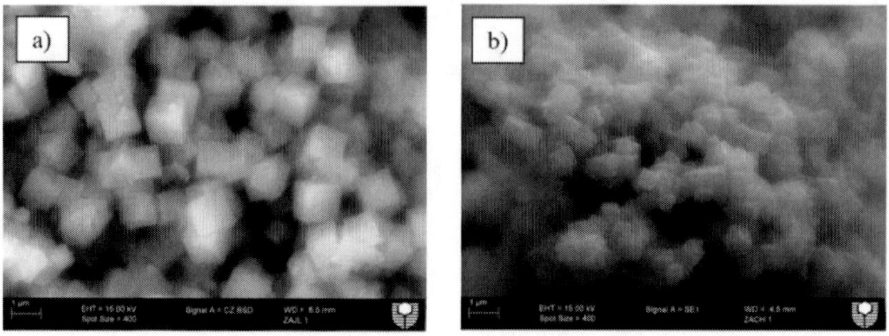

Figure 2. SEM micrographs of a) ZAJL; b) ZACH.

3.1.3. Specific Surface Area

Table 1 shows the specific surface area for the pure zeolites. Zeolites A presented 15.7 m^2 g^{-1} and 10.9 m^2 g^{-1} for ZAJL and ZACH, respectively. The specific surface area for the pure zeolites is much higher than those presented by the raw fly ashes or the products synthesized by the conventional hydrothermal treatment using the same samples (Izidoro et al., 2012).

3.1.4. Cation Exchange Capacity

According to Table 1, the value of CEC for both pure zeolites was 3.9 meq g^{-1}. This value is in accordance with the values found in literature (Hollman et al., 1999; Yaping et al., 2008). The CEC values presented by the pure zeolites are much higher than those of the untreated material or the products synthesized using one-step hydrothermal treatment (Izidoro et al., 2012).

Table 1. Characteristics of zeolites A from Brazilian fly ashes

Sample	**S_{BET} (m^2 g^{-1})**	**CEC (meq g^{-1})**
ZAJL	15.7	3.9
ZACH	10.9	3.9

3.2. Adsorption Efficiencies in Single and Binary Systems

Initially, the capacity of the pure zeolites to remove Cd^{2+} and Zn^{2+} ions from single aqueous solutions was tested at various concentrations. The adsorption percentages for both ions are shown in Table 2.

According to Table 2, the pure zeolites synthesized from Brazilian fly ashes were found to successfully remove cadmium and zinc ions from single aqueous solution at high concentrations. The metal removal percentages for both ions decreased with increasing concentrations due to the adsorbent saturation.

In the single systems, the adsorption percentage of zeolites A ranged from 59.6 to 100.0% and from 97.2 to 100.0% for Cd^{2+} and Zn^{2+} ions, respectively. The similarity of the pore size of zeolite A (0.42 nm) with the diameter of the Cd^{2+} and Zn^{2+} hydrated ions (Zn^{2+} presents hydrated ion size = 0.430 nm and Cd^{2+} presents hydrated ion size = 0.426 nm) may have contributed to the removal of those ions (Breck, 1974; Chang & Shih, 2000).

It was also found that the amount of adsorbed Zn^{2+} was larger than that of Cd^{2+} on both zeolites. Similar results have been reported in adsorption studies (Srivastava et al., 2008).

This preference of Zn^{2+} to pure zeolites may be due to the loss of water molecules of hydration by the ions during the adsorption process. The hydrated ion size of Cd^{2+} and Zn^{2+} is very similar but during the adsorption process some water molecules are lost, the ion size for both ions reduces and the size for Zn^{2+} is 0.083 nm, smaller than that of Cd^{2+} (0.103 nm) (Semmens & Seyfarth, 1981). Therefore, ions having a smaller radius are better than those with larger hydrated radius to fill in the pores in adsorbent surface. Furthermore, the atomic weight of Zn^{2+} is lower than that presented by Cd^{2+}. Lighter ions may be more easily adsorbed than heavy ions, resulting in a selectivity order related to the atomic weight by the zeolites.

Table 2a. Comparison of individual and total adsorption percentages by pure zeolites found at different Cd^{2+} and Zn^{2+} concentrations after 24 h of contact

Adsorbent	$C_{o,Cd}$(mg L^{-1})	$C_{o,Zn}$ (mg L^{-1})	Ad_{Cd}%	Ad_{Zn}%	Ad_{Tot}%
ZAJL (Single)	1124.1	0.0	100.0	-	100.0
	1348.9	0.0	91.8	-	91.8
	1686.2	0.0	74.8	-	74.8
	2248.2	0.0	79.3	-	79.3
	2810.3	0.0	60.0	-	60.0
	3372.3	0.0	59.6	-	59.6
	0.0	653.8	-	100.0	100.0
	0.0	784.6	-	98.7	98.7
	0.0	980.7	-	98.9	98.9
	0.0	1307.6	-	99.1	99.1
	0.0	1634.5	-	97.7	97.7
	0.0	1961.4	-	97.3	97.3
ZAJL (Binary)	1124.1	653.8	92.4	100.0	95.2
	1348.9	653.8	90.6	100.0	93.7
	1686.2	653.8	74.2	100.0	81.4
	2248.2	653.8	73.5	88.5	76.9
	2810.3	653.8	57.4	73.1	60.4
	3372.3	653.8	54.8	61.9	55.9
	1124.1	784.6	97.2	90.7	94.5
	1124.1	980.7	84.4	88.7	86.4
	1124.1	1307.6	76.9	72.1	74.3
	1124.1	1634.5	67.5	60.0	63.1
	1124.1	1961.4	58.7	46.8	51.1
ZACH (Single)	1124.1	0.0	100.0	-	100.0
	1348.9	0.0	92.4	-	92.4
	1686.2	0.0	92.9	-	92.9
	2248.2	0.0	79.8	-	79.8
	2810.3	0.0	60.7	-	60.7
	3372.3	0.0	60.0	-	60.0
	0.0	653.8	-	100.0	100.0
	0.0	784.6	-	99.2	99.2
	0.0	980.7	-	99.3	99.3
	0.0	1307.6	-	99.2	99.2
	0.0	1634.5	-	97.6	97.6
	0.0	1961.4	-	97.2	97.2

Table 2b. Comparison of individual and total adsorption percentages by pure zeolites found at different Cd^{2+} and Zn^{2+} concentrations after 24 h of contact

Adsorbent	$C_{o,Cd}$(mg L^{-1})	$C_{o,Zn}$ (mg L^{-1})	Ad_{Cd}%	Ad_{Zn}%	Ad_{Tot}%
ZACH (Binary)	1124.1	653.8	93.1	100.0	95.6
	1348.9	653.8	91.4	100.0	94.2
	1686.2	653.8	90.0	100.0	92.8
	2248.2	653.8	77.0	82.3	78.1
	2810.3	653.8	60.2	66.4	61.4
	3372.3	653.8	59.7	57.3	59.3
	1124.1	784.6	100.0	86.9	94.6
	1124.1	980.7	94.5	81.5	88.4
	1124.1	1307.6	81.9	75.3	78.3
	1124.1	1634.5	72.7	61.7	66.2
	1124.1	1961.4	66.7	51.9	57.3

The pH values for the supernatant solutions after the adsorption processes in single systems were measured. The average pH for the Cd^{2+} and Zn^{2+} solutions were 6.6 and 6.8, respectively. All pH values were lower than the precipitation pH for Cd^{2+} (which precipitates at pH around 8.4) and Zn^{2+} (which precipitates at pH values between 7.0 and 8.0).

In the binary studies, the influence of competitive cations was tested by varying the initial concentration of one metal ion while the initial concentration of the other competitor cation was fixed. Table 2 summarizes the adsorption percentages for Cd^{2+} and Zn^{2+} under different initial concentrations of the competitor cation.

According to Table 2, the adsorption percentages in the binary systems were in general lower than those in the single systems due to the competition between Cd^{2+} and Zn^{2+} ions to occupy exchangeable places in the zeolites. Thus, zinc solutions under the presented conditions reduce the uptake of cadmium and vice versa.

The total metal ion adsorption percentages (Ad_{Tot} %) for zeolite A ranged from 51.1 to 95.6%, respectively. By comparing the Ad_{Tot}% of Cd^{2+} and Zn^{2+} ions between single and binary systems, it is noted that zeolites A still reached high adsorption percentages with the presence of the competitor ions. In general, cadmium removal was more affected by the presence of zinc than the opposite due to the preference for Zn^{2+} ions by the pure zeolites.

The pHs of all binary solutions after the adsorption process were under the precipitation pH for both studied ions. The average pH for all solutions was 6.2.

3.3. Adsorption Isotherms in Single Systems

The adsorption equilibrium data for single systems using different adsorbents were evaluated by Langmuir and Freundlich models. These two isotherm equations have been used extensively by several authors to evaluate the equilibrium characteristics of adsorption processes.

In the Lagmuir model, the basic assumption is that the sorption occurs at specific homogeneous sites within the adsorbent, whereas in the Freundlich model, it is considered that the adsorbent surface is heterogeneous and the adsorption process occurs at a non-uniform distribution of heat over the surface (Tran et al., 2017).

Figures 3 and 4 show the adsorption isotherms for Cd^{2+} and Zn^{2+}, respectively, with the Langmuir and Freundlich models. The isotherm parameters determined by the Langmuir and Freundlich equations are shown in Tables 3 and 4 for Cd^{2+} and Zn^{2+}, respectively.

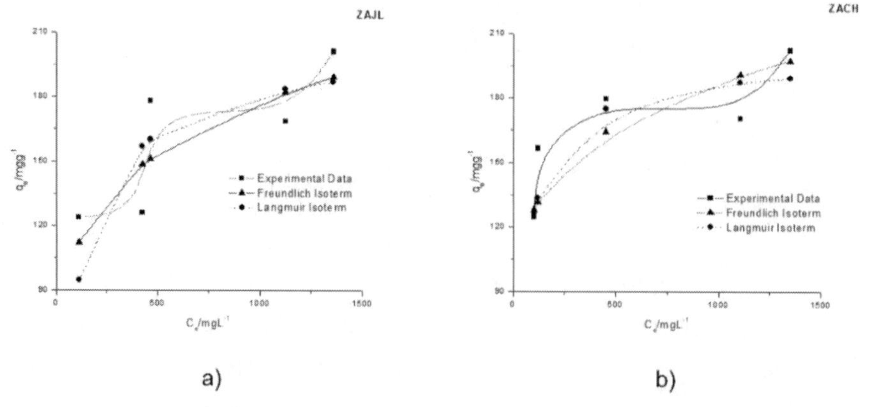

Figure 3. Adsorption isotherms of Cd^{2+} on a) ZAJL; b) ZACH (T = 298.15 K).

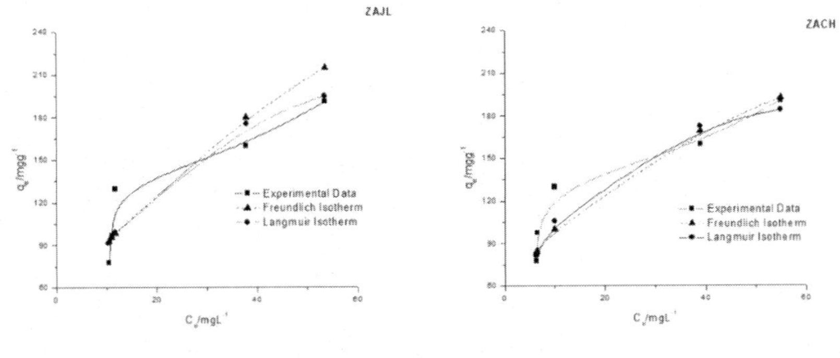

a) b)

Figure 4. Adsorption isotherms of Zn^{2+} on a) ZAJL; b) ZACH (T = 298.15 K).

Table 3. Langmuir and Freundlich parameters calculated from Cd^{2+} isotherms in single-ion system using pure zeolites synthesized from fly ash

Adsorbent	Langmuir Parameters		
	Q_0 (mg g^{-1})	K_L (L mg^{-1})	R
ZAJL	185.1	0.0178	0.959
ZACH	181.2	0.0587	0.967
Adsorbent	**Freundlich Parameters**		
	K_f^*	n	R
ZAJL	51.4	5.55	0.969
ZACH	78.3	7.99	0.982

(*)(mg g^{-1}) (L mg^{-1})$^{1/n}$.

Table 4. Langmuir and Freundlich parameters calculated from Zn^{2+} isotherms in single-ion system using pure zeolites synthesized from fly ash

Adsorbent	Langmuir Parameters		
	Q_0 (mg g^{-1})	K_L (L mg^{-1})	R
ZAJL	219.6	0.0857	0.957
ZACH	212.7	0.157	0.967
Adsorbent	**Freundlich Parameters**		
	K_f^*	n	R
ZAJL	41.7	2.65	0.974
ZACH	54.9	3.28	0.983

(*)(mg g^{-1}) (L mg^{-1})$^{1/n}$.

According to the correlation coefficients (R) showed in Tables 3 and 4, the experimental data for all isotherms complied with the Langmuir and Freundlich models.

As can be observed, the adsorption capacities (Qo) for Cd^{2+} were very close for the two samples (185.1 and 181.2 mg g^{-1}, for ZAJL and ZACH, respectively). The zinc adsorption capacities were higher than that of cadmium and also showed very similar values for the two zeolitic materials (Table 4). The stronger affinity to zinc than cadmium on all pure zeolites was also demonstrated by higher values of adsorption energy, K_L, (Table 3 and 4), which is a constant related to the affinity between an adsorbent and adsorbate

3.4. Adsorption Isotherms in Binary Systems

The adsorption equilibrium data for the binary studies were evaluated using the parameters calculated by the Langmuir and Freundlich linear equations only for a comparison with the existing data for the monocomponent systems. The isotherm parameters determined for Cd^{2+} and Zn^{2+} ions are shown in Tables 5 and 6, respectively.

Table 5. Langmuir and Freundlich parameters calculated from Cd^{2+} isotherms in binary-ion system using pure zeolites synthesized from fly ash

Adsorbent	Langmuir Parameters		
	Q_o (mg g^{-1})	K_L (L mg^{-1})	R
ZAJL	88.7	2.74	0.969
ZACH	90.1	2.59	0.940
Adsorbent	**Freundlich Parameters**		
	K_f^*	n	R
ZAJL	63.0	16.5	0.987
ZACH	58.1	12.9	0.972

(*)(mg g^{-1}) (L mg^{-1})$^{1/n}$.

Table 6. Langmuir and Freundlich parameters calculated from Zn^{2+} isotherms in binary-ion system using pure zeolites synthesized from fly ash

Adsorbent	Langmuir Parameters		
	Q_o (mg g^{-1})	K_L (L mg^{-1})	R
ZAJL	174.6	0.0160	0.974
ZACH	195.3	0.0158	0.989
Adsorbent	**Freundlich Parameters**		
	K_f^*	n	R
ZAJL	48.4	5.64	0.984
ZACH	55.8	5.81	0.982

(*)(mg g^{-1}) (L $mg^{-1})^{1/n}$.

In the same way as in the single-ion systems, the correlation coefficients (R) showed that the adsorption process could be described by both Langmuir and Freundlich models (Tables 5 and 6). The adsorption capacities in the binary (Q_o) for all adsorbents were lower than those values in the single systems (Tables 3 and 4) due to the competition between Cd^{2+} and Zn^{2+} ions.

4. Zeolite A from Fly Ash and Aluminum Waste

4.1. Synthesis and Characterization

4.1.1. Characterizations of Aluminum Waste

The chemical compositions of RA and others physicochemical properties are shown in Table 7. The major component was Al_2O_3, Cl and Na_2O along with percentages range between 3.8 and 1.6 wt% of SiO_2, MgO, K_2O and FeO_2.

The final waste from the tertiary industry consists of metallic aluminum, aluminum oxide and a large amount of salts used as fluxing material. Their chemical composition varies according to the type and quality of the processed scrap, the classification and trapping methods used, etc. The high Cl and Na content are attributable to the employ of high amounts of salt (mainly NaCl and KCl) in the melting process to get a higher aluminum recovery (Shinzato and Hypolito, 2005). Others metals may be associated with the different kinds of scrap used in secondary industry.

The pH value indicates that RA was moderately alkaline in nature due the presence of aluminum and magnesium oxide, which hydrolyze in the presence of water. The conductivity of the leachate can be attributed to the presence of large amounts of soluble salts in the residue, such as NaCl, KCl, etc.

The soluble aluminium content refers to all the aluminium compounds which dissolve in hydrochloric acid (10% HCl for a reaction time of 3 hr at 80 °C), such as, metallic aluminium, aluminium nitride, sulfide and different types of salts. The total aluminium content determined for XFR also refers to other insoluble compounds.

The XRD pattern of the residue of aluminum tertiary industry is shown in Figure 5. The crystalline phases were mainly composed of corundum (Al_2O_3; ICDD 42-1468), sylvite (KCl; ICDD 0041-1476), spinel ($MgAl_2O_4$; ICDD 002-1084), halite (NaCl; ICDD 001-0994) and fayalite ($2FeO.SiO_2$; ICDD 002-0784). The high background of the XRD profile also indicates the presence of non-well crystallized or amorphous phases in which metallic oxides such as iron oxide, among others, could be included.

Table 7. Physicochemical properties of RA

Parameters	Values
Al_2O_3 (wt. %)	27.8
Cl (wt. %)	17.4
Na_2O (wt. %)	16.7
SiO_2 (wt. %)	3.78
MgO (wt. %)	2.37
K_2O (wt. %)	2.27
Fe_2O_3 (wt. %)	1.59
Others (wt. %)	≤ 0.9
$Al_{soluble}$ (wt. %)	16.2
pH	7.6
Conductivity (mS cm^{-1})	22.8

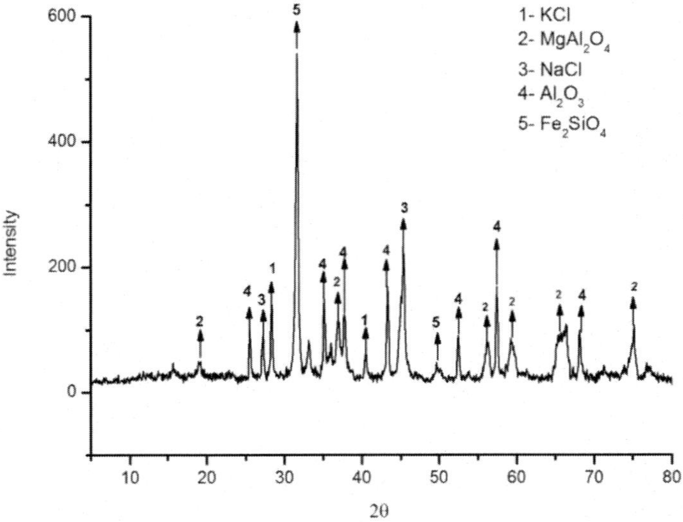

Figure 5. XRD pattern of the aluminum waste sample.

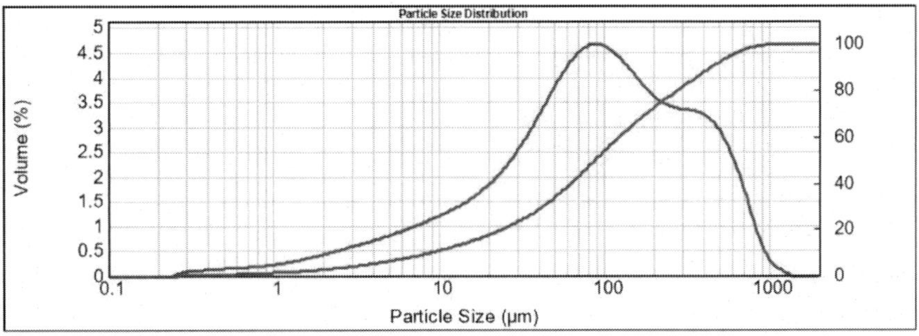

Figure 6. Particle size distribution of RA.

Figure 6 reports the differential and cumulative particle size distributions for RA. The material tends to show a bimodal particle size distribution ranging from 0.283 μm to 1261.915 μm.

The particle size distributions of material are given in Table 8. Cumulative size distributions calculated at 90% (D_{90}) indicates high particle diameters (~450 μm). So, this sample presents high percentage of coarse particles indicating that was collected by a worm screw system.

Table 8. Particle size distributions of material

	Particle diameter (μm)
D_{10}	9.088
D_{50}	89.404
D_{90}	448.715
D([3,2])	15.414

Table 9. Concentration of elements leachate from RA and the allowed limit values

Element	**Leachate (mg L^{-1})**	**Allowed limit (mg L^{-1})**
Arsenic	ND*	1.0
Barium	3.57	70.0
Cadmium	0.013	0.5
Chromium	0.016	5.0
Lead	0.149	1.0
Mercury	ND*	0.1
Selenium	ND*	1.0
Silver	< 0.010	5.0

(*)ND-data not available.

Table 10. Concentration of elements solubilised from RA and the allowed limit values

Element	**Leachate (mg L^{-1})**	**Allowed limit (mg L^{-1})**
Aluminum	2,560	0.2
Arsenic	ND*	0.01
Barium	0.28	0.7
Cadmium	ND*	0.005
Chromium	< 0.010	0.05
Copper	0.349	2.0
Iron	7.6	0.3
Lead	ND*	0.01
Manganese	< 0.010	0.1
Mercury	ND*	0.001
Selenium	ND*	0.01
Sodium	ND*	200
Silver	0.033	0.05
Zinc	0.104	5.0

(*)ND-data not available.

RA sample was submitted to acetic acid leaching and water solubilization tests, according to Brazilian regulation. The results were shown in Table 9 and 10, respectively, with the limit values determined by the Brazilian Standard NBR 10004 (ABNT NBR 20004, 2004).

The high concentration of aluminum in the extract leachate (> 26,547 mg L^{-1}) interfered in the determination of As, Se and Hg (Table 9). Table 10 showed that concentrations of Al and Fe were above the maximum limit allowed, according to the Brazilian Standard NBR 10004 and the interference of high concentration of solubilized aluminum does not allow the determination of the concentration of some elements.

4.1.2. Characterization of Synthesized Zeolites

The effect of the fusion temperature and time on the zeolitization process was evaluated and experimental parameters are presented in Table 11. X-ray diffraction analysis was used to identify the crystalline phases of synthesized zeolites from aluminum waste and fly ash (Figure 7). The results indicated that the all products contain zeolite A as the major constituent phase, whereas hydroxysodalite, quartz and mullite were found as minor phases.

Table 11 also shows the characterization of the products in *terms* of cation exchange capacity (CEC) and an estimate of zeolite content in the synthesized material. The CEC was used to estimate semi-quantitative data of the zeolite content in the synthesized material by comparing the CEC value obtained with pure commercial zeolite A (4.55 meq g^{-1}). This estimate considers that CEC values for minor phases are negligible (Querol, X.; Moreno, N.; Umaña, J. C.; Alastuey, A.; Hernández, E.; López-Soler, A.; Plana, F.; Int. J. Coal Geol. 2002, 50, 413).

The CEC value ranged from 3.10 to 3.77 meq g^{-1}. The highest CEC was observed for fusion temperature of 600ºC and 2 h of fusion time. However, the difference is not significant for both zeolite percentages and CEC values. Thus, the condition used to obtain ZRA-1 was selected in further studies because is the most economical condition.

Table 11. Experimental conditions of fusion treatment and characterization of zeolites from residue of aluminum and fly ash

Zeolitic Material	T (°C)	Time (h)	CEC(1) (meq g^{-1})	Content (%)
ZRA-1	550	1	3.34	73.4
ZRA-2	550	2	3.10	68.1
ZRA-3	600	1	3.38	74.3
ZRA-4	600	2	3.77	82.9

(1) cation-exchange capacity.

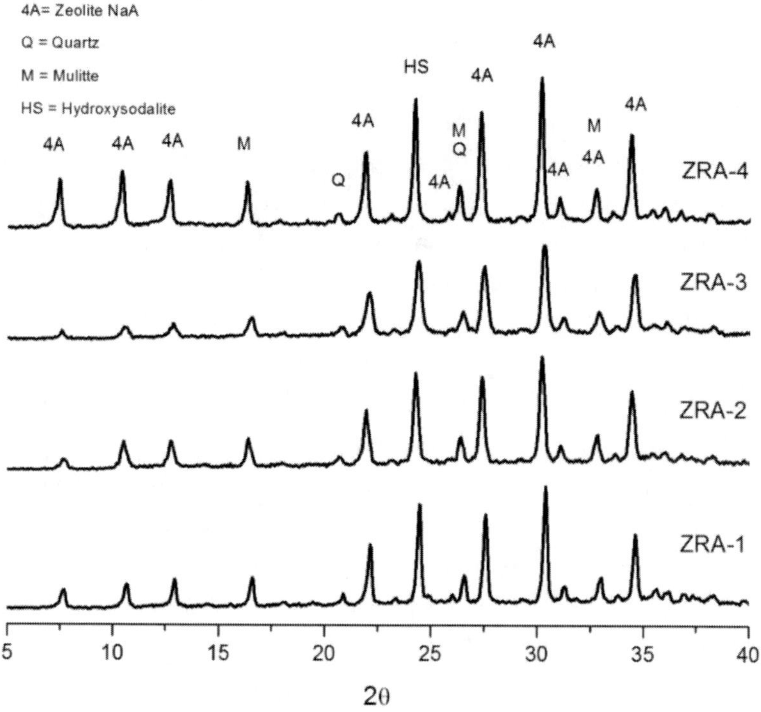

Figure 7. XRD patterns of zeolites synthesized from aluminum waste and coal fly ash.

The formation of the zeolite A is not effective when the residue was added in the aging step after the fusion step only with fly ash and NaOH, taking into account that the synthesis product contained a substantial percentage of hydroxysodalite and with some quartz. *The mineralogy of RA* predominantly consisted of crystalline phases of aluminum. Thus, a melting step is necessary so that the aluminum that the residue contains dissolves and reacts with the silicon present in the ashes to form the sodium aluminosilicate sufficient to form Na-A zeolite.

Other Al-containing wastes were used in the synthesis of zeolite A. Zeolite NaA could be synthesized using Si extracted from rice husk waste, and aluminium cans, as a source of Al, applying different procedures for the preparation of initial hydrogel and a unified procedure for crystallization of zeolite (Shoumkova and Stoyanova, 2013; Ahmedzeki et al. 2018).

Two industrial wastes, the solid by-product of cupola slag and aluminum sludge from an aluminum plating plant, were successfully converted into Na-A zeolite by alkaline fusion treatment. The optimum precursor was the mixture of treated cupola slag: NaOH: aluminum sludge in the ratio of 1:3:1 at 700ºC for 1 h (Anuwattana and Khummongkol, 2009).

Zeolite A production was also possible utilizing by-product from aluminium etching (Nugteren, 2010; Kanokorn et al. 2011).

4.2. Adsorption Studies

The adsorption equilibrium data of Cd(II) and Zn(II) onto ZRA-1 were analyzed by Langmuir, Freundlich, Temkin, Dubinin-Radushkevich (D-R) models. Table 12 shows the values of isotherm parameters and the adjustable determination coefficients (R^2). The modeled isotherms for Cd^{2+} and Zn^{2+} are plotted in Figure 8 and 9, respectively.

As shown in the Table 12, all the used models correlated well with the experimental data ($R^2 > 0.9$) for both metal–zeolite adsorption systems, except for the Langmuir model for Zn.

The values of error analysis for Langmuir, Freundlich, Temkin and D-R isotherm models are presented in Table 13. As shown in Table 12 and Table 13, the D-R adsorption isotherm model yielded best fit to the experimental equilibrium for Cd(II) adsorption according to the R^2 and error values. The higher values of R^2 and lower values of error functions indicated that the Freundlich model was more suitable for describing the adsorption equilibrium of Zn(II) onto the zeolite adsorbent.

Table 12. Parameters of the Langmuir, Freundlich, Temkin and D-R models for adsorption of Cd^{2+} and Zn^{2+} onto ZRA-1

Parameters	Cd^{2+}	Zn^{2+}
	Langmuir	
Q_o (mg g^{-1})	78.0	35.8
b_L (L mg $^{-1}$)	0.265	0.254
R^2	0.935	0.752
	Freundlich	
k_F (mg g^{-1})(L mg^{-1})$^{1/n}$	40.0	13.4
1/n	0.112	0.223
R^2	0.945	0.920
	Temkin	
k_T (L g^{-1})	143.0	8.94
B_T	7.13	5.32
b_t (kJ mol^{-1})	0.348	0.466
R^2	0.970	0.899
	D-R	
β (mol^2 J^{-2})	1.20 x 10^{-9}	2.20 x 10^{-9}
k_{DR} (mol g^{-1})	8.79 x 10^{-4}	9.95 x 10^{-4}
E (kJ mol^{-1})	20.4	15.1
R^2	0.973	0.921

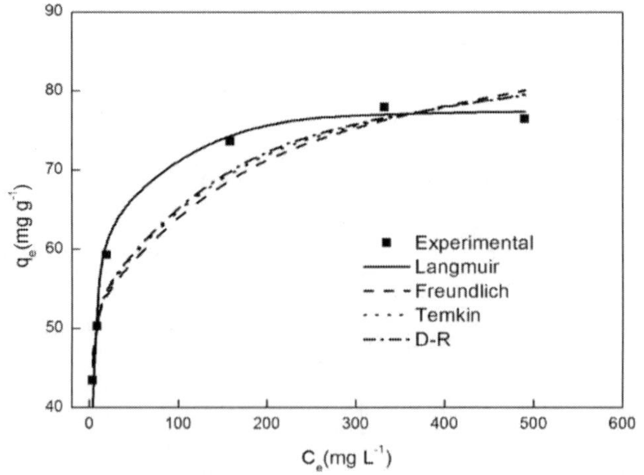

Figure 8. Adsorption isotherms of Cd^{2+} on ZRA-1.

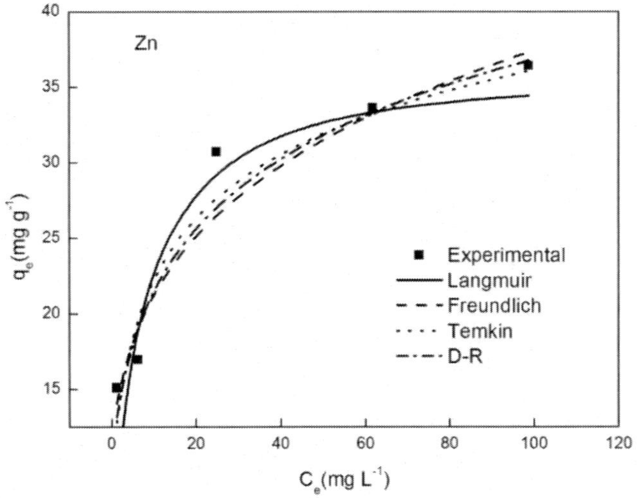

Figure 9. Adsorption isotherms of Zn^{2+} on ZRA-1.

Table 13. Error functions analysis related to the adsorption of adsorption of Cd^{2+} and Zn^{2+} onto ZRA

Cd^{2+}	ARED	SSE	MPSD	HYBRID	SAE	X^2
Langmuir	5.38	73.7	57.9	8.05	18.0	1.28
Freundlich	4.21	46.9	43.9	6.3	15.3	0.766
Temkin	3.01	26.0	31.7	4.50	11.3	0.403
D-R	2.97	23.6	30.8	4.45	10.9	0.376
Zn^{2+}	ARED	SSE	MPSD	HYBRID	SAE	X^2
Langmuir	13.3	72.2	105.4	19.9	13.8	6.14
Freundlich	6.53	23.3	51.7	9.78	8.46	1.01
Temkin	8.18	29.5	64.6	12.2	9.10	1.48
D-R	7.22	23.3	55.0	10.8	8.51	1.13

CONCLUSION

Two methods have been evaluated for the synthesis of zeolite A from coal fly ash using alkaline fusion step followed by hydrothermal treatment and different aluminum sources.

In the first method, pure zeolites A were synthesized from two Brazilian fly ashes with different chemical compositions. The ashes were used without pre-treatment and no seeds or structure-directing agents were added in the synthesis, which occurred without extraction step. The formation of Zeolite A was achieved by the addition of commercial sodium aluminate in the aging step as aluminum source.

The cubic characteristic shape of zeolite A appeared on SEM images in conformity with the successful formation of the well-organized zeolitic framework.

The two synthesized materials were used for the treatment of water containing high levels of Zn^{2+} and Cd^{2+} ions in single or binary systems. Higher affinity for Zn^{2+} than Cd^{2+} for both single and binary systems was observed under similar experimental conditions. The adsorption percentages in the binary systems were in general lower than those presented for the single systems.

The adsorption process could be described by both Langmuir and Freundlich models for all isotherms in single and binary systems.

In the second method, Na-A zeolite from coal fly ash has been successfully prepared using aluminum waste from tertiary industry as aluminum source. The X-ray diffraction analysis demonstrated that NaA was the main zeolitic phase formed after the synthesis achieving a yield between 68 and 83%.

The prepared zeolite was tested to remove zinc and cadmium ions from water at high concentration. The effectiveness of adsorption reached 79-99% for Zn^{2+} and 61-99% for Cd^{2+}.

The D-R and Freundlich model was the most appropriate for fit of the equilibrium experimental data for Cd and Zn, respectively.

Zeolites A synthesized by addition of commercial sodium aluminate presented cation exchange capacity value close to that aluminum waste-based zeolite.

It can be concluded that the use of commercial sodium aluminate or aluminum residue as aluminum source was efficient in the production of zeolite A from fly ash using an alkaline fusion process followed by a hydrothermal treatment. Moreover, synthesized materials can be used to remove metal ions during water treatment process.

ACKNOWLEDGMENTS

The authors are grateful to Jorge Lacerda and Charqueadas Coal-fired Power Plants and Company Latasa Recycling for providing samples. The support of Capes Foundation (Ministry of Education of Brazil) and Curtin University.

REFERENCES

ABAL - Associação Brasileira Do Alumínio. (2017). *Brazilian aluminum: solutions for sustainable living*. ABAL, São Paulo.

ABNT NBR 10004 - Associação Brasileira De Normas Técnicas. (2004). *NBR 10004: Solid Waste e Classification*. ABNT, Rio de Janeiro.

ABNT NBR 10005 - Associação Brasileira De Normas Técnicas. (2004). *NBR 10005: Leaching Tests*. ABNT, Rio De Janeiro.

ABNT NBR 10006 - Associação Brasileira De Normas Técnicas. (2004). *NBR 10006: Solubilization Tests*. ABNT, Rio De Janeiro.

Ahmedzeki, N. S., Abbas, M. N., Joodee, A. M., Jaed, Y. M. (2018). Waste Resources Utilization for Zeolite a Synthesis. *J. Chem. Technol. Metal.*, 53(2): 239-244.

Alcântara, R. R., Muniz, R. O. R, Fungaro, D. A. (2016). Full factorial experimental design analysis of Rhodamine B removal from water using organozeolite from coal bottom ash. *Intern. J. Energy Environ.*, 7(5): 357-374.

Anuwattana, R., Khummongkol, P. (2009). *Conventional hydrothermal synthesis of Na-A zeolite from cupola slag and aluminum sludge. J. Haz. Mat.*, 166: 227-232.

Breck, D. W. (1974). *Zeolite Molecular Sieves: Structure, Chemistry, and Use*. Wiley, New York.

Chang, H. L., Shih, W. H. (2000). Synthesis of zeolites A and X from fly ashes and their ion-exchange behavior with cobalt ions. *Ind. Eng. Chem. Res.*, 39: 4185-4191.

Chareonpanich, M., Jullaphan, O., Tang, C. (2011). Bench-scale synthesis of zeolite A from subbituminous coal ashes with high crystalline silica content. *J. Clean. Prod.*, 19: 58-63.

El-Naggar, M. R., El-Kamash, A. M., El-Dessouky, M. I., Ghonaim, A. K. (2008). Two-step method for preparation of NaA-X zeolite blend from fly ash for removal of cesium ions. *J Hazard. Mater.*, 154: 963-972.

Fansuri, H., Yanti, Y., Londar, E., Widiastuti, N. (2010). Synthesis of Zeolite-A from Coal Bottom Ash: The Influence of Unburned carbon to the Zeolite Quality. In: *The 13th Asia Pacific Confederation of Chemical Engineering Congress*, APCChE 2010, Taipei, Taiwan.

Fungaro, D. A., Izidoro, J. C., Santos, F. S., Wang, S. (2013). Coal Fly Ash from Brazilian Power Plants: Chemical and Physical Properties and Leaching Characteristics. In: Sarker, P. K. (2013). *Fly Ash: Chemical Composition, Sources and Potential Environmental Impacts.* Nova Science Publishers, New York.

Gil, A., Korili, S. A. (2016). Management and valorization of aluminium saline slags: current status and future trends, *Chem. Eng. J.*, 289:74-84.

Hollman, G. G., Steenbruggen, G., Jansen-Jurkovicova, M. (1999). A two-step process for the synthesis of zeolites from coal fly ash. *Fuel*, 78: 1225-1230.

Hong, J. L. X., Maneerung, T., Koh, S. N. N., Kawi, S., Wang, C. H. (2017). Conversion of Coal Fly Ash into Zeolite Materials: Synthesis and Characterizations, Process Design, and Its Cost-Benefit Analysis. *Ind. Eng. Chem. Res.*, 56:11565-11574.

Hui, K. S., Chao, C. Y. H. (2006). Effects of step-change of synthesis temperature on synthesis of zeolite 4A from coal fly ash. *Micropor. Mesopor. Mater.*, 6188: 145-151.

Izidoro, J. C., Fungaro, D. A., Abbott, J. E., Wang, S. (2013). Synthesis of zeolites X and A from fly ashes for cadmium and zinc removal from aqueous solutions in single and binary ion systems. *Fuel*. 103: 827-834.

Izidoro, J. C., Fungaro, D. A., Santos, F. S., Wang, S. (2012). Characteristics of Brazilian coal fly ashes and their synthesized zeolites, *Fuel Process. Technol.*, 97: 38-44.

Izidoro, J. C., Fungaro, D. A., Abbott, J. E., Wang, S. (2013). Synthesis of zeolites X and A from fly ashes for cadmium and zinc removal from aqueous solutions in single and binary ion systems. *Fuel Process. Technol.*, 103: 827-834.

Kanokorn, H., Sombat, T., and Nisakorn, S. (2011). Synthesis of zeolite A from by-product *of* aluminum etching process: Effects of reaction temperature and reaction time on pore volume. *Am. J. Environ. Sci.*, 7(1): 35–42.

Koshy, N., Singh, D. N. (2016). Fly ash zeolites for water treatment applications. *J. Environ. Chem. Eng.*, 4:1460-1472.

López-Delgado, A., Rodríguez, O., Padilla, I., Galindo, R., López-Andrés, S. (2014). Industrial aluminum hazardous waste as a new raw material for zeolite synthesis. In: *WIT Transactions on Ecology and the Environment*, WIT Press, United Kingdom.

López-Delgado, A., Tayibi, H. (2012). Can hazardous waste become a raw material? The case study of an aluminum residue: a review. *Waste Manage. Res.*, 30: 474-484.

Mallapur, V. P., Oubagaranadin, J. U. K. (2017). A Brief Review on the Synthesis of Zeolites from Hazardous Wastes. *T. Indian Ceram. Soc.*, 76(1): 1-13.

Nugteren, H. W. (2010). *Secondary industrial minerals from coal fly ash and aluminium anodising waste solutions.* [PhD thesis]. Delft University of Technology.

Ojumu, T. V., Du Plessis, P. W., Petrik, L. F. (2016). Synthesis of zeolite A from coal fly ash using ultrasonic treatment - A replacement for fusion step. *Ultrason. Sonochem.*, 31: 342-349.

Panitchakarn, P., Laosiripojana, N., Viriya-umpikul, N., Pavasant, P. (2014) Synthesis of high-purity Na-A and Na-X zeolite from coal fly ash, *J. Air Waste Manage. Assoc.*, 64(5): 586-596.

Petrov, I., Michalev, T. (2012). Synthesis of Zeolite A: A Review. *Научни Трудове На Русенския Университет*, 51: 30-35. [*Scientific Work of the Russian University*]

Querol, X., Moreno, N., Umaña, J. C., Alastuey, A., Hernandez, E., Lopez-Soler, A. (2002). Synthesis of zeolites from coal fly ash: an overview. *Int J Coal Geol.*, 50: 413–423.

Ríos, R. C. A., Williams, C. D., Roberts, C. L. (2009). A comparative study of two methods for the synthesis of fly ash-based sodium and potassium type zeolites. *Fuel.* 88: 1403–1416.

Sánchez-Hernández, R., López-Delgado, A., Padilla, I., Galindo, R., and López-Andrés, S. (2016). *One-step synthesis of NaP1, SOD and ANA from a hazardous aluminum solid waste. Microp. Mesop. Mater.*, 226: 267-277.

Sánchez-Hernández, R., Padilla, I., López-Andrés, S., López-Delgado, A. (2017). *Eco-friendly bench-scale zeolitization of an Al-containing waste* into gismondine-type zeolite under effluent recycling. *J. Clean. Prod.*, 161: 792-802.

Semmens, M. J., Seyfarth, M. (1981). The selectivity of clinoptilolite for certain heavy metals. In: Sand, L. B. and Mumpton, F. A. *Natural zeolites: occurrence, properties, and use*. Pergamon, Oxford.

Shigemoto, N., Hayashi, H. (1993). Selective formation of Na-X zeolite from coal fly ash by fusion with sodium hydroxide prior to hydrothermal reaction, *J. Mater. Sci.*, 28: 4781-4786.

Shinzato, M. C., Hypolito, R. (2005). Solid Waste from aluminum recycling process: characterization and reuse of its economically valuable constituents. *Waste Manage.*, 25: 37-46.

Shinzato, M. C., Hypolito, R. (2012). Effect of disposal of aluminum recycling waste in soil and water bodies. *Environ. Earth Sci.*, 75 (7): 1-10.

Shoumkova, A., Stoyanova, V. (2013). SEM–EDX and XRD characterization of zeolite NaA, synthesized from rice husk and aluminum scrap by different procedures for preparation of the initial hydrogel. *J. Porous Mater., 20: 249-255.*

Somerset, V., Petrik, L., Iwuoha, E. (2007). Alkaline hydrothermal conversion of fly ash precipitates into zeolites 3: the removal of mercury and lead ions from wastewater. *J. Environ. Manage.*, 87: 125-131.

Srivastava, V. C., Mall, I. D., Mishra, I. M. (2008). Removal of cadmium(II) and zinc(II) metal ions from binary aqueous solution by rice husk ash. *Colloids and Surfaces A: Physicochem. Eng. Aspects*, 312: 172-184.

Tanaka, H., Fujii, A. (2009). Effect of stirring on the dissolution of coal fly ash and synthesis of pure-form Na-A and –X zeolites by two-step process. *Adv. Powder Technol*, 20: 473-479.

Tran, H. N., You, S. J., Hosseini-Bandegharaei, A., Chao, H. P. (2017). Mistakes and inconsistencies regarding adsorption of contaminants from aqueous solutions: A critical review. *Water Res.,* 120: 88-116.

Tsakiridis, P. E. (2012). Aluminium Salt Slag Characterization and Utilization - A Review. *J. Haz. Mat.*, 217-218: 1-10.

Volli, V., Purkait, M. K. (2015). Selective preparation of zeolite X and A from fly ash and its use as catalyst for biodiesel production. *J. Hazard. Mater.*, 297: 101-111.

Wang, C. F., Li, J. S., Wang, L. J., Sun, X. Y. (2008). Influence of NaOH concentrations on synthesis of pure-form zeolite A from fly ash using two-stage method, *J. Hazard. Mater.*, 155: 58-64.

Yaping, Y., Xiaoqiang, Z., Weilan, Q., Mingwen, W. (2008). Synthesis of pure zeolites from supersaturated silicon and aluminum alkali extracts from fused coal fly ash. *Fuel*, 87: 1880-1886.

Ye, Y., Zeng, X., Qian, W., Wang, M. (2008). Synthesis of pure zeolites from supersaturated silicon and aluminum alkali extracts from fused coal fly ash. *Fuel*. 87: 1880-1886.

INDEX

A

B

C

D

E

F

G

H

I

L

M

N

O

P

Fly Ash: Properties, Analysis and Performance

Editor: Jacob Parker

Series: Materials Science and Technologies

Book Description: Coal fly ash (CFA) is one of the most complex anthropogenic materials. It is estimated that only about 20 to 30% of the globally generated fly ash is employed and utilized in building materials mainly as an additive in cement, concrete, and structural filling as well as in small scale production of zeolite. This book provides new research on the characteristics, uses and performance of fly ash.

Hardcover ISBN: 978-1-53610-516-2
Retail Price: $195

Fly Ash: Sources, Applications and Potential Environmental Impacts

Editor: Prabir Kumar Sarker (Curtin University, Australia)

Series: Environmental Science, Engineering and Technology

Book Description: Fly ash is a by-product of the combustion of coal and other waste materials. This book presents some of the latest developments in the generation, characterization, utilization and environmental impacts of fly ash.

Hardcover ISBN: 978-1-62948-044-2
Retail Price: $230